A-Z
GENERAL MICROBIOLOGY

A-Z GENERAL MICROBIOLOGY

Prof. Nirmal Chandra Pradhan

CENTRUM PRESS
NEW DELHI-110002 (INDIA)

CENTRUM PRESS

H.O.: 4360/4, Ansari Road, Daryaganj,
New Delhi-110 002 (India)
Ph.: 23278000, 23261597

B.O.: No. 1015, Ist Main Road, BSK IIIrd Stage
IIIrd Phase, IIIrd Block,
Bangalore - 560 085 (India)
Tel.: 080-41723429
Visit us at: www.centrumpress.com

A-Z General Microbiology

First Edition, 2009

ISBN 978-93-80106-27-4

PRINTED IN INDIA

Printed at Salasar Imaging Systems, Delhi-110035 (India)

Contents

Preface

Microbiology is the study of *microorganisms,* which are unicellular or cell-cluster microscopic organisms. This includes eukaryotes such as Fungi and Protists, and prokaryotes, which are Bacteria and Archaea. Viruses, though not strictly classed as living organisms, are also studied. In short; microbiology refers to the study of life and organisms that are too small to be seen with the naked eye.

Microbiology is a broad term which includes Virology, Mycology, Parasitology, Bacteriology and other branches. A microbiologist is a specialist in microbiology. Microbiology is researched actively, and the field is advancing continually. We have probably only studied about one percent of all of the microbe species on Earth. Although microbes were first observed over three hundred years ago, the field of microbiology can be said to be in its infancy relative to older biological disciplines such as Zoology and Botany.

Beneficial roles of microbes include recycling of organic matter through microbe-induced decay and through digestion and nutrition in animals and humans. In addition, the natural microbial flora provides protection against more virulent microbes. While microbes that cause infectious diseases are virulent, opportunistic diseases may also be caused by normally benign microbes. Opportunistic infections occur when the host defense mechanisms are impaired; microbes are present in large numbers, or when microbes reach vulnerable body sites.

Author

Chapter 1

Microbiology

Microbiology can be defined as the study of organisms too small to be seen with the naked eye. shows the relative size of microbes compared to other living things. However, the recent discovery of bacteria of near 1 mm in size has made this definition somewhat inaccurate and in the grand tradition of science, a new definition is in order. Microbes for being small, span a large range of sizes from the T4 virus virus that are 0.02 μm to giant bacteria larger than 700 μm.

We will consider microbiology to be the study of organisms that can exist as single cells, contain a nucleic acid genome for at least some part of their life cycle, and are capable of replicating that genome. This definition would also include viruses, which microbiology texts traditionally discuss along with living organisms. Microbiology also involves a collection of techniques to study and manipulate these small creatures. Because of their size, special instruments and methods had to be developed to allow the performance of interpretable experiments on microorganisms.

These methods are not restricted to microbes alone, but have also found utility in working with populations of cells from higher organisms.Many different organisms fall under the definition of microorganisms. Shown here are; A, the microorganism *Escherichia coli*; B, a photosynthetic cyanobacterium (Mike Clayton); C, a fungus (CDC); D, Ebola virus (CDC); E, the malaria parasite (a protozoan) (CDC).

Microbiology also involves a collection of techniques to study and manipulate these small creatures. Because of their size, special instruments and methods had to be developed to allow the performance of interpretable experiments on microorganisms. These methods are not restricted to microbes alone, but have also found utility in working with populations of cells from higher organisms.

HUMAN EFFECTED BY MICROBES

"Ancient" diseases continue to be a problem where nutrition and sanitation are poor, and emerging diseases such as Acquired Immunodeficiency Syndrome (AIDS) are even more dangerous for such populations. The Centers for Disease Control and Prevention (the U.S. government agency charged with protecting the health and safety of people) estimates that about 9% of adults between the ages of 18-49 in Sub-Saharan Africa are infected with HIV. AIDS is only one of a number of new diseases that have emerged.

Many of these diseases have no known cure. Influenza and pneumonia are leading killers of the elderly even in the U. S. and other developed nations. Even the common cold causes illness and misery for almost everyone and drains the productivity of all nations. Many of the new diseases are viral in nature, making them notoriously difficult to treat. Disease due to food-borne pathogens also remains a problem, largely because of consumption of improperly processed or stored foods.

Understanding the sources of contamination and developing ways to limit the growth of pathogens in food is the job of food microbiologists. New infections continually appear.Having an available food source to grow on (humans) inevitably results in a microorganism that will take advantage. Some of these feeders will interfere with our own well being, causing disease. Surprisingly, many diseases that were previously thought to have only behavioral or genetic components have been found to involve microorganisms. The clearest case is that of ulcers, which was long thought to be caused by stress and poor diet. However the causative agent is actually a bacterium, *Helicobacter pylori*, and many ulcers can be cured with appropriate antibiotics.

Work on other non-infectious diseases such as heart

disease, stroke and some autoimmune diseases also suggest a microbial component that triggers the illness. Finally, some pathogenic microbes that had been "controlled" through the use of antibiotics are beginning to develop drug resistance and therefore reemerge as serious threats in the industrialized world as well as developing nations. Tuberculosis is an illness that was on the decline until the middle 80's. It has recently become more of a problem, partly due to drug resistance and partly due to a higher population of immunosuppressed individuals from the AIDS epidemic.

Staphylococcus aureus strains are emerging that are resistant to many of the antibiotics that were previously effective against them. These staph infections are of great concern in hospital settings around the world. Understanding both familiar killers and new pathogens will require an understanding of their biology, and thus an understanding of the field of microbiology.

BENIFITS OF MICROBES

Significant resources have been spent to understand and fight disease-causing microorganisms from the beginning of microbiology. You may be surprised to learn that only a small faction of microbes are involved in disease; many other microbes actually enhance our well-being. The harmless microbes that live in our intestines and on our skin actually help us fight off disease. They actively antagonize other bacteria and take up space, preventing potential pathogens from gaining a foothold on our bodies.

Indeed, like all large organisms, humans have entire communities of microorganisms in their digestive systems that contribute to their overall health. The microbial community in humans not only protects us from disease, but also provides needed vitamins, such as B_{12}. Human health and nutrition also depends on healthy farm animals. Cows, sheep, horses and other ruminant animals utilize their microbial associates to degrade plant material into useful nutrients.

Without these bacteria inside ruminants, growth on plant material would be impossible. In contrast to humans ruminant animals have complex stomachs that harbor large numbers of microorganisms. These microbes degrade the plant stuff eaten by the animal into usable nutrients. Without the assistance of the microbes, ruminant animals would not be able to digest the food they eat. Commercial crops are also central to human prosperity, and much of agriculture depends upon the activities of microbes. For example, an entire group of plants, the legumes, forms a cooperative relationship with certain bacteria.

These bacteria convert nitrogen gas to ammonia for the plant, an important nutrient that is often limiting in the environment. Microbes also serve as small factories, producing valuable products such as cheese, yogurt, beer, wine, organic acids and many other items. In conclusion, while it is less apparent to us, the positive role of microbes in human health is at least as important as the negative impact of pathogens. A picture of nodulated leguminous plants. In this case pea plants. Nodules are visible on the roots of the plants in the left of the picture. The plants on the right were not inoculated with nodulating bacteria.

ENVIRONMENT EFFECTED BY MICROBES

The vast majority of life on this planet is microscopic. These teaming multitudes profoundly influence the make-up and character of the environment in which we live. Presently, we know very little about the microbes that live in the world around us because less than 2 % of them can be grown in the laboratory. Understanding which microbes are in each ecological niche and what they are doing there is critical for our understanding of the world. Microbes are the major actors in the synthesis and degradation of all sorts of important molecules in environments.

Cyanobacteria and algae in the oceans are responsible for the majority of photosynthesis on Earth. They are the ultimate source of food for most ocean creatures (including whales) and replenish the world's oxygen supply. Cyanobacteria also use carbon dioxide to synthesize all of their biological molecules and thus remove it from the atmosphere. Since carbon dioxide is a major greenhouse gas, its removal by cyanobacteria affects the global carbon dioxide balance and may be an important mitigating factor in global warming. In all habitats, microorganisms make nutrients available for the future growth of other living things by degrading dead organisms.

Microbes are also essential in treating the large volume of sewage and wastewater produced by metropolitan areas, recycling it into clean water that can be safely discharged into the environment.

Less helpfully (from the view of most humans), termites contain microorganisms in their guts that assist in the digestion of wood, allowing the termites to extract nutrients from what would otherwise be indigestible. Understanding of these systems helps us to manage them responsibly and as we learn more we will become ever more effective stewards. Energy is essential for our industrial society and microbes are important players in its production.

A significant portion of natural gas comes from the past action of methanogens (methane-producing bacteria). Numerous bacteria are also capable of rapidly degrading oil in the presence of air and special precautions have to be taken during the drilling, transport and storage of oil to minimize their impact. In the future, microbes may find utility in the direct production of energy. For example, many landfills and sewage treatment plants capture the methane produced by methanogens to power turbines that produce electricity.

Excess grain, crop waste and animal waste can be used as nutrients for microbes that ferment this biomass into methanol or ethanol. These biofuels are presently added to gasoline and thus decreasing pollution. They may one-day power fuel cells in our cars, causing little pollution and having water as their only emission.

Finally, We are increasingly taking advantage of the versatile appetite of bacteria to clean up environments that we have contaminated with crude oil, polycholrinated biphenyls (PCBs) and many other industrial wastes. This process is termed bioremediation and is a cheap and increasingly effective way of cleaning up pollution.

MICROBES AROUND US

Microorganisms will grow on simple, cheap medium and will often rise to large populations in a matter of 24 hours. It is easy to isolate their genomic material, manipulate it in the test tube and then place it back into the microbe. Due to their large populations it is possible to identify rare events and then, with the use of powerful selective techniques, isolate interesting bacterial cells and study them. These advantages have made it possible to test hypothesis rapidly.

Using microbes scientists have expanded our knowledge about life. Below are a few examples. Microorganisms have been indispensable instruments for unlocking the secrets of life. The molecular basis of heredity and how this is expressed as proteins was described through work on microorganisms. Due to the similarity of life at the molecular level, this understanding has helped us to learn about all organisms, including ourselves. Some prokaryotes are capable of growing under unimaginably harsh conditions and define the extreme limits of where life can exist. Some species have been found growing at near 100 °C in hot springs and well above that temperature near deep-sea ocean vents.

Others make their living at near 0 °C in freshwater lakes that are buried under the ice of Antarctica. The ability of microbes to live under such extreme conditions is forcing scientists to rethink the requirements necessary to support life. Many now believe it is entirely possible that Jupiter's moon Europa may harbor living communities in waters deep below its icy crust. What may the rest of the universe hold? Until recently, while we could study specific types of bacteria, we lacked a cohesive classification system, so that we could not readily predict the properties of one species based on the known properties of others.

Visual appearance, which is the basis for classification of large organisms, simply does not work with many microbes because there are few distinguishing characteristics for comparison between species even under the microscope. However, analysis of their genetic material in the past 20 years has allowed such classification and spawned a revolution in our thinking about the evolution of bacteria and all other species. The emergence of a new system organizing life on Earth into three domains is attributable to this pioneering work with microorganisms. The fruits of basic research on microbes has been used by scientists to understand microbial activity and therefore to shape our modern world.

Human proteins, especially hormones like insulin and human growth factor, are now produced in bacteria using genetic engineering. Our understanding of the immune system was developed using microbes as tools. Microorganisms also play a role in treating disease and keeping people healthy. Many of the drugs available to treat infectious disease originate from bacteria and fungi. One last recent role of microbes in informing us about our world has been the tools they provide for molecular biology. Enzymes purified from bacterial strains are useful as tools to perform many types of analyses. Such analyses allow us to determine the complete genome sequence of almost any organism and manipulate that DNA in useful ways.

We now know the entire sequence of the human genome, with the exception of regions of repetitive DNA, and this will hopefully lead to medical practices and treatments that improve health. We also know the entire genome sequence of many important pathogens. Analysis of this data will eventually lead to an understanding of the function of critical enzymes in these microbes and the development of tailor-made drugs to stop them. The tools of molecular biology will also affect agriculture.

For example, we now know the complete genome sequence of the plant *Arabidopsis* (a close relative of broccoli and cauliflower). This opens a new avenue to a better understanding of all plants and hopefully improvements in important crops. Microbes have a profound impact on every facet of human life and everything around us. Pathogens harm us, yet other microbes protect us. Some microbes are pivotal in the growth of food crops, but others can kill the plants or spoil the produce.

Bacteria and fungi eliminate the wastes produced in the

environment, but also degrade things we would rather preserve. Clearly they effect many things we find important as humans. In the remainder of this chapter we take a look at how scientists came to be interested in microbes and follow a few important developments in the history of microbiology.

THE HISTORY OF MICROBIOLOGY

A look at the history of microbiology will help you to understand the contributions of those who have come before. This perspective will hopefully give you an appreciation of their efforts and put the body of knowledge we will examine in the context of history. Keep in mind that microbiology is a relatively young science. It was only 130 years ago that it became possible to seriously study microorganisms in the laboratory, with most of our understanding of microbes coming in the last 60 years. The history of microbiology, like all human history, is not a catalog of linear progress, but is more of an interweaving of the careers of bright individuals and their insights.

Each new discovery relied on previous ones and in turn spawned further inquiry. A web of interdependent concepts evolved over time through the work of scientists in many related disciplines and nations. Often the research of one individual impacted the efforts of another studying a completely different problem. Keep this in mind as you look at this history. Below we present several journeys through this web, mentioning some individuals who were particularly important in the progression. This history reflects our view of important events of the past, but is by no means comprehensive.

We will first look at the development of the techniques for handling microorganisms, since everything else in microbiology depends upon these procedures. Next, we will examine how these techniques helped to settle an old debate, the question of spontaneous generation. Then, we will look at the history of infectious disease. The science of microbiology had its most significant early impact on human health, uncovering the cause of the major killers of the day, and then methods to treat them. As microbiology matured, scientists began to look at what non-pathogenic microbes were doing in the environment and we will look a bit at the history of general microbiology. Finally, the chapter will end with an examination of the events that lead to the understanding of life at the molecular level and the profound impact this has had on microbiology and on society in general.

IMPORTANCE OF MICROSCOPES

It took the microscope to expose their tiny world and that instrument has been linked to microbiology ever since. In 1664, Robert Hooke devised a compound microscope and used it to observe fleas, sponges, bird feathers, plants and molds, among other items. His work was published in *Micrographia* and became a popular and widely read book at the time. Several years later Anton van Leeuwenhoek, a fabric merchant and amateur scientist (or "natural philosopher" as such people called themselves), became very adept at grinding glass lenses to make telescopes and microscopes. While crude by modern standards, his were a technical marvel for the time, able to magnify samples greater than 200-fold.

They also produced clearer images than the compound microscopes of the time. By peering through his microscope, Leeuwenhoek observed tiny organisms or "wee animacules" as he called them.. He spent months looking at every kind of sample he could find and eventually submitted his observations in a letter to the Royal Society of London, causing a sensation. Hooke was asked to confirm the findings of Leeuwenhoek and his affirmative assessment garnered them wide acceptance. Surprisingly the work of these two scientists was not followed up for almost 200 years.

Human societies had neither the technical prowess nor the inclination to develop the science of microorganisms. It was not until the rise of the industrial revolution that governments and people dedicated the financial and physical resources to understand these small inhabitants of our world. With the development of better microscopes in the 19th Century, scientists returned to an examination of microorganisms. After finishing his education, Ferdinand Julius Cohn was able to convince his father to lay down the large sum necessary to purchase a microscope for him, one better than that available at the University in Breslau, then part of Germany.

He used it to carefully examine the world of the microbe and made many observations of eukaryotic microorganisms and bacteria. His landmark papers on the cycling of elements in nature was published in *Ueber Bakterien* in 1872 and a microbial classification scheme including descriptions of *Bacillus* were published in the first volume of a journal he founded, *Beitraege zur Biologie der Planzen*. Cohn's work with microscopes popularized their use in microbiology. This and his other work inspired many other scientists to examine microbes. Cohn's encouragement of Robert Koch, a German physician by training, began the field of medical microbiology.

MICROBIOLOGY TECHNIQUES

A major contribution to bacterial techniques was the development of methods using solid medium for the cultivation of bacteria. Koch was convinced that microbes caused some diseases. However, to test this idea, he needed to isolate the causative agent. Almost all samples from diseased animals or any natural surface contained many different microbes and it was impossible to tell which one was the problem. A method was needed to separate these different bacteria.

The most common method of isolation was to continually dilute a sample in liquid broth in hopes that at high enough dilution, only one type of microbe would be found. A problem with this method is that only the most populous microbe would be isolated, but that might not be one causing the disease. There were other technical problems as well with such a liquid-based system, so a solid medium would seem to provide distinct advantages. Koch had tried gelatin for these experiments with unsatisfactory results.

Building on the work of Brefeld and Schroeter, Koch used potato slices as a solid medium and observed that a boiled potato left in the open air would develop tiny circular raised spots. Examination of these spots revealed they were made up of microorganisms and each spot had just one type of microbe in it. He realized that these colonies were pure cultures of bacteria and probably arose from a single species of microbe from the air that landed on the potato. By boiling a potato, slicing it with a hot knife and keeping it in a sterile container with a lid, Koch could keep the potato sterile.

But if a sample from a disease animal was smeared across the potato, colonies arose, each being pure isolates from the animal. By then testing these isolates in animals, Koch was able to isolate the cause of anthrax, *Bacillus anthracis*. Potatoes failed to support the growth of many microorganisms and Koch and his laboratory were constantly frustrated by the lack of a good solid medium. Walter Hesse joined Koch's laboratory to do studies on air quality, showing a remarkable attention to detail and patience in his work.

His wife, Angelina Fannie Hesse along with raising their three sons, also would assist her husband with his research in the laboratory. Walter was attempting to do his air quality experiments using medium containing gelatin as the solidifying agent. In the summertime, temperatures would often rise above the melting point of gelatin, then used as the solidifying agent. In addition, microbes would often grow in the cultures that were capable of degrading gelatin and in both cases this would cause liquefaction of the medium, ruining the experiments.

One day while eating lunch, the frustrated scientist asked Lina (as she was called) why her jellies and puddings stayed solid even in the hot summer temperatures. She told him about agar-agar, a heat resistant gelling agent that she had learned about while growing up in New York from a Dutch neighbour who had emigrated from Java. Development of the new agent by Angelina and Walter led to a resounding success. Few microbes are able to degrade agar and it melts at 100°C yet remains molten at temperatures above 45°C. This allows the mixing of the agar with heat-sensitive nutrients and microbes.

After solidification, it will not melt until a temperature of 100°C is again attained, facilitating the easy cultivation of pathogens. It can also be stored for long periods of time, allowing the cultivation of slow-growing microorganisms. Any type of broth can be mixed with agar, giving great flexibility in the kinds of medium that can be made. Thus, many more types of microbes could be cultivated. Koch's laboratory also developed methods of pure culture maintenance and aseptic technique. Aseptic technique involves the manipulation of pure cultures in a manner that prevents their contamination by outside microorganisms.

Equally important, aseptic technique prevents their spread into the environment. Remember that Koch was studying some of the most devastating microbial pathogens of the period, and their release could potentially cause disease in the scientists working on them. These procedures were also absolutely critical because they allowed careful study of pure microorganisms, making it possible to identify the role of each microbe in a given situation.

Another problem in the cultivation of microbes was solved by Julius Petri while working in Koch's laboratory. Solid medium was poured on glass plates and allowed to spread and harden. Once cooled it allowed a solid surface for streaking. However, creation of these plates required great care since exposure to the air (and the microbes in it) often lead to contamination.

In addition, to prevent contamination of plates during incubation, a cumbersome bell jar was used. If one wanted to view samples, the plate had to be removed from the jar, further exposing it to unwanted microbes in the air. In 1887 Petri developed shallow glass dishes, with one having a slightly larger diameter than the other. Medium is poured into the smaller dish and the larger one serves as a cover. This simple device solved all of the above problems and took on the name of its inventor, the petri plate.These same techniques are essential in studying all microorganisms. Collectively the above techniques have been used to isolate and identify thousands of different microorganisms.

As a testament to the significance of their achievement, these techniques are practiced with remarkably little change in every laboratory that works with microorganisms today.Spontaneous generation is the hypothesis that some vital force contained in or given to organic matter can create living organisms from inanimate objects. Spontaneous generation was a widely held belief throughout the middle ages and into the latter half of the 19th century. In fact, some people still believe in it today. The idea was attractive because it meshed nicely with the prevailing religious views of how God created the universe.

There was a strong bias to legitimize the idea because this vital force was considered a strong proof of God's presence in the world. Many recipes and experiments were offered in proof. To create mice, a recipe called for dirty underwear and wheat grain to be mixed in a bucket and left open outside. In 21 days or less, you would have mice. The real cause may seem obvious from a modern perspective, but to the proponents of this idea, the mice spontaneously arose from the wheat kernels. Another often-used example was the generation of maggots from meat that was left in the open. The failing here was revealed by Francesco Redi in 1668 with a classic experiment. Redi suspected that flies landing on the meat laid eggs that eventually grew into maggots.

Here he used three pieces of meat.One piece of meat was placed under a piece of paper. The flies could not lay eggs onto the meat and no maggots developed. The second piece was left in the open air, resulting in maggots. In the final test, a third piece of meat was overlaid with cheesecloth. The flies were able to lay the eggs into the cheesecloth and when this was removed no maggots developed. However, if the cheesecloth containing the eggs was placed on a fresh piece of meat, maggots developed, showing it was the eggs that "caused" flies and not spontaneous generation.

This helped to end the debate about spontaneous generation for large organisms. However, spontaneous generation was so seductive a concept that even Redi believed it was possible in other circumstances. Using several pieces of meat, paper and cheesecloth, Francesco Redi demonstrated strong evidence against the theory of spontaneous generation.

One of the strong points of this experiment was its simplicity, allowing others to easy test it for themselves. The concept and the debate were revived in 1745 by the experiments of John Needham. It was known at the time that heat was lethal to living organisms.

Needham theorized that if he took chicken broth and heated it, all living things in it would die. After heating some broth, he let a flask cool and sit at a constant temperature. The development of a thick turbid solution of microorganisms in the flask was strong proof to Needham of the existence of spontaneous generation. Lazzaro Spallanzani later repeated the experiments of Needham, but removed air from the flask, suspecting that the air was providing a source of contamination.

No growth occurred in Spallanzani's flasks and he took this as evidence that Needham was wrong. Proponents of spontaneous generation discounted the experiment by asserting that air was required for the vital force to work. It was not until almost 100 years later that the great French chemist Louis Pasteur put the debate to rest. He first showed that the air is full of microorganisms by passing air through gun cotton filters. The filter trapped tiny particles floating in the air. By dissolving the cotton with a mixture of ether and alcohol, the particles were released and then settled to the bottom of the liquid.

Inspection of this material revealed numerous microbes that resembled the types of bacteria often found in putrefying media. Pasteur realized that if these bacteria were present in the air then they would likely land on and contaminate any material exposed to it. Pasteur then entered a contest sponsored by The French Academy of Sciences to disprove the theory of spontaneous generation. Similar to Spallanzani's experiments, Pasteur experiment used heat to kill the microbes, but left the end of the flask open to the air.

In a simple, but brilliant modification, the neck of the flask was heated to melting and drawn out into a long S-shaped curve, preventing the dust particles and their load of microbes from ever reaching the flask. After prolonged incubation the flasks remained free of life and ended the debate for most scientists. Pasteur filled a flask with medium, heated it to kill all life, and then drew out the neck of the flask into a long S. This prevented microorganisms in the air from entering the flask, yet allowed air to flow freely. If the swan neck was broken, microbes could enter the flask and grow. A final footnote on the topic was added when John Tyndall showed the existence of heat-resistant spores in many materials.

Boiling does not kill these spores and their presence in chicken broth, as well as many other materials, explains the results of Needham's experiments. While this debate may seem silly from a modern perspective, remember that the scientists of the time had little knowledge of microorganisms. Koch would not isolate microbes until 1881. The proponents of spontaneous generation were neither sloppy experimenters nor stupid. They did careful experiments and interpreted them with their own biases.

Detractors of the theory of spontaneous generation were just as guilty of bias, but in the opposite direction. In fact, it is somewhat surprising that Pasteur and Spallanzoni did not get growth in their cultures, since the sterilization conditions they used would often not kill endospores. Luck certainly played a role. It is important keep in mind that the discipline of science is performed by humans with all the fallibility and bias inherent in the species. Only the self-correcting nature of the practice reduces the impact of these biases on generally held theories.

Spontaneous generation was a severe test of scientific experimentation, because it was such a seductive and widely held belief. Yet, even spontaneous generation was overthrown when the weight of careful experimentation argued against it.

MICROBIAL CAUSE OF DISEASE

Living things were the agents of disease. This was only one idea among many and some thought that an imbalance in humors caused illness, while others felt that supernatural forces were at work. The prevailing theory held by most doctors of the 19th century was that chemical toxins were carried from an ill patient to others, causing them to contract the same malady. The bacteria that were known to be present were seen as a symptom of the disease and not its cause. Ignaz Semmelweis a Hungarian physician working in Vienna, made the first breakthrough in the true nature of disease. He realized that asepsis in obstetrical wards could prevent the transmission of childbirth fever from patient to patient.

He therefore instigated a policy for all attending physicians to wash their hands with with chloride of lime (a mixture of calcium chloride hypochlorite, CaCl(OCl); calcium hypochlorite, $Ca(OCl)_2$; and calcium chloride, $CaCl_2$) between patients. This innovation dropped the mortality rate from 18% to 2.4%. Ignaz became a vigorous proponent of his ideas, but the Hungarian doctor's efforts were opposed by many who could not accept that physicians themselves could be responsible for spreading bacterial infection.

Ridicule of his idea caused him to move from Vienna to Pest, Hungary and ultimately played a role in a nervous breakdown. Ironically Semmelweis died from an infection that he contracted during a surgery he performed, while recovering from his nervous breakdown. Before his death he published his ideas in a paper The *Cause, Concept, and Prophylaxis of Childbed Fever* in 1861. Although the poor writing of the paper contributed to the obscurity of his ideas, the work was ignored for 17 years, which raises an interesting point about the culture of science. Radical ideas, even those that are correct and can save lives, are sometimes ignored.

It takes time to overcome the dogma of the day. The personalities involved and the negative light it might throw on past practices play a large role in the rate of acceptance of a new idea. In a seemingly unrelated event, Louis Pasteur found something interesting while working on wine souring, a problem where wine fermentations produce a sour taste and very little alcohol. Fermentation of alcoholic beverages was thought to be a simple chemical reaction. Heating, common in most beverage preparations, was thought to cause the breakdown of sugar into alcohol. Pasteur was asked to help out the wine industry in France because wine souring was pushing it close to ruin. His work on wine-making revealed that the process of converting sugar to alcohol is actually performed by various yeast strains.

He then showed that the wine was going bad because a contaminating microbe was generating lactic acid instead of alcohol from the sugar. This idea was controversial, but gained credibility when Pasteur solved the problem by heating the wine and killing the contaminant. The heating process was named pasteurization in his honour and is still widely used today. In a brilliant step of generalization, Pasteur realized that souring of wine and infectious disease shared a common thread in that they both might involve infection by a microorganism. His suggestion that microbes cause disease became known as the germ theory of disease. Joseph Lister became aware of Semmelweis's work and together with Pasteur realized the true nature of disease. He then recognized that he could use this idea to help his surgery patients.

At this time, major injuries, broken bones or surgery would often result in infection of the damaged area, sometimes leading to amputation or death. Lister found he could greatly reduce the number of microorganisms on wounds and incisions by using bandages treated with phenic acid, a compound that killed microorganisms (phenic acid, now known as phenol, is the active ingredient in Listerine). During surgery he began the practice of spraying the wound with a fine mist of phenic acid to kill microbes.

These practices greatly reduced the rate of infection and mortality of surgery patients, lending further credence to the germ theory of disease. In 1876 Robert Koch provided definitive proof of the germ theory by isolating the cause of anthrax and showing it to be a bacterium. From this came the development of Koch's Postulates, a set of rules for the assignment of a microbe as the cause of a disease:

- The specific organism should be shown to be present in all cases of animals suffering from a specific disease, but should not be found in healthy animals.
- The specific microorganism should be isolated from

the diseased animal and grown in pure culture on artificial laboratory media.

- This freshly isolated microorganism, when inoculated into a healthy nonimmune laboratory animal, should cause the same disease seen in the original animal.
- The microorganism should be reisolated in pure culture from the experimental infection.

The postulates are Koch's most famous contribution to science and it is a testament to the utility of these postulates that they are stilled used today to discover the cause of new emerging diseases. Koch went on to apply these principles in the study of many other diseases including tuberculosis, cholera and sleeping sickness. It should be pointed out that Koch's postulates cannot be applied to all diseases.

For example if a disease-causing microbe has humans as its sole host and has a significant possibility of causing death, it would be unethical to apply this microbe to test humans as dictated by postulate 3. Also, it is not always possible to obtain a disease-causing microbe in pure culture. While attacking the problem of disease, Koch developed the tools for obtaining pure cultures. Advances in science often come from innovations in the available technology. Robert Koch was an important microbiologist because his pioneering work in the isolation and characterization of bacterial diseases helped to identify the causes of many of the maladies plaguing humanity. Further work by other scientists then began the long road to conquering them.

VIRUSES

In the latter half of the 19th century, the causative agents for anthrax, tuberculosis, gonnorhea, diptheria and many more were discovered. Great strides in understanding maladies caused by bacteria were made at this time, including the solidification of the germ theory of disease. However, some illnesses seemed not to have bacterial origins. Microscopic examination of sera from ill patients revealed no organisms and the causative agents could not be grown on any known medium. Yet if these sera were injected into a susceptible host, disease resulted. In 1886 John Brown Buist devised a method for staining and fixing liquid from a cowpox vesicle. Observations of this slide showed tiny bodies that he believed were spores.

Although he did not realise it, he was the first person to see (and photograph) a virus. It was not until the middle of the twentieth century with the invention of the electron microscope that the true shape and structure of viruses was understood. In 1884 Charles Chamberland in Pasteur's laboratory created an unglazed porcelain filter that had pores much smaller than bacteria (0.1-1 μm). It was possible to pass a solution containing bacteria through these filters and completely remove them from the solution. This enabled the creation of sterile medium without the use of heat, and also became a standard test for the removal of all bacteria, especially when testing transmission of disease.

By removing the microbes, Chamberland clearly demonstrated that an infectious agent, and not the solution, was causing the illness. This concept seemed reasonable until scientists began to investigate a tobacco infection. Tobacco mosaic disease is characterized by light and dark green areas on plant leaves in a mosaic pattern. The disease stunts the growth of the plant, therefore reducing yields. Europeans recognized the effects of this disease soon after tobacco was introduced from the New World in the 17th century. Adolph Mayer first described transmission of the disease by injecting fluid from a diseased plant into a healthy one.

Nine out of ten times the healthy plant would become heavily diseased. In 1892 Dmitrii Ivanowski extended this research by the shocking discovery that the causative agent could pass through Chamberland's porcelain filter. He reported his findings, but the idea that the causative agent could pass through the filter was so troublesome to Ivanowsky he attributed the phenomenon to a cracked filter or to small spores that passed through the pores. It was Martinus Beijerinck in 1898 who realized the true nature of these particles, making the intellectual leap that the causative agent of the disease must be so small as to pass through the filter known to trap all bacteria. He coined the term *contagium vivum fluidum* a contagious living fluid. Some scientists thought these agents were toxins in the fluids of the hosts, but further study revealed that tobacco mosaic disease could cause illness through many transfers from plant to plant.

A toxin might cause damage on the first plant, but subsequent transfers should make it so dilute that it would no longer have any effect. These filterable entities were different from bacteria and appeared to depend on their host in order to multiply. The term ultrafiltrable viruses and later just viruses (virus means poison in Latin) was coined to describe these tiny pathogens. Bacteria are also vulnerable to viruses. Bacterial viruses were described independently by Frederick Twort in 1915 and Felix d'Herelle in 1917. Felix d'Herelle was studying a plague of locusts in Mexico when he noticed that the locusts were being killed by a microorganism.

During experiments to characterize this microbe, d'Herelle noticed the appearance of clear, circular spots two or three millimeters in diameter on cultures growing on agar. At this point d'Herelle studied the spots enough to determine that they came from an agent small enough to pass through a porcelain filter. He dropped the investigation, but recalled these observations while studying dysentery in 1915. The illness was affecting soldiers fighting in World War I and he soon determined the cause to be the bacterium *Shigella dysentery*. In the process of investigating *Shigella*, similar areas of clearing were observed. d'Herelle realized that something in these areas of clearing was killing the bacterium. He eventually determined these were viruses of bacteria, coining the term bacteriophage.

VACCINATION

The search was on for ways to kill or to prevent them from causing disease. In this section and the next, we will look at two series of events that illustrate the emergence of modern treatment of infectious disease: the development of vaccines and the discovery of antimicrobial compounds.

Smallpox was a feared disease throughout human history and justifiably so. It was highly contagious and almost everyone eventually became infected. Mortality rates were as high as 25% in adults and closer to 40% in children.

Those who did survive often had scarring due to the blister-like pustules that form on the skin, but they obtained life-long immunity to the disease. As far back at the 11th century in India and China it was realized that liquid from the pustules of a smallpox victim, when scratched on the skin of a healthy patient, would most often cause mild disease. This intentional infection, termed variolation, would also give life-long protection against the virus. Lady Mary Wortley Montgue, wife of ambassador to the Ottoman Empire, introduced variolation to England in 1721 and it became a popular practice throughout Europe.

Washington even began variolating the Continental Army in 1776. Variolation had some deleterious side effects. Serious skin lesions inevitably resulted at the site of inoculation, often accompanied by a generalized rash or even a full case of smallpox. The fatality rate from variolation was 1 to 2 %. Today we would find this level of fatality to be unacceptable, but at the time this risk still represented a significant advance. In 1796, Edward Jenner went in search of a more predictable and safer method of protection against the disease. He noticed that milkmaids rarely contracted smallpox.

Further investigation revealed they often contracted cowpox from their charges. Jenner hypothesized that cowpox was related to smallpox and contraction of the former would protect against the latter. In a classic experiment (and one that would land you in jail today), Jenner inoculated a young patient with cowpox and later challenged him with smallpox. The boy did not become ill and Jenner was responsible for the creation of a safer method of protection against smallpox. It is important to stress that the nature of these diseases and their viruses would not be known for over 100 years. Jenner was ahead of his time. Beginning around 1876 Pasteur's studies on chicken cholera led to the development of vaccines to fight the disease. Cholera was a serious problem since it was able to spread through a barnyard and wipe out a flock in as little as 3 days. It was transmitted by contaminated food or animal excrement.

Pasteur identified the cholera bacillus and grew it in pure culture. When injected with it, a chicken invariably died within 48 hours. Then, as often happens in scientific research, luck intervened. During the heat of the summer, Pasteur returned to Paris and left the cholera cultures used for infection stored on the shelves of his laboratory in Arbois, France. Upon returning, something had happened to the cultures, they no longer caused disease when tested in chickens.

With some impatience for the time they were wasting, his group set to work making new cultures of the bacillus and tested these batches on both new birds and also those previously inoculated with the ineffective strain. To their amazement the previously injected birds were unaffected by the fresh bacillus culture, while the new birds all died. Pasteur immediately realized that this was similar to the studies of Jenner.

Pasteur then developed a method for creating cultures that would confer immunity, but not cause disease. Sometimes this involved growing the microbe in medium in the laboratory where they would spontaneously loose their virulence. In other cases it involved multiple passes through a susceptible host. For example, the rabies virus was attenuated by passing it through rabbits.

In honour of Jenner's accomplishments Pasteur coined the term vaccination (*vacca* = cow in Latin) for the process of immunization against disease. In this and several of Pasteur's other discoveries, luck played a part, but it was only helpful because he tenaciously pursued "odd" results and had the insight to arrive at important conclusions.

In Pasteur's famous words, " In the field of observation, chance favors only the prepared mind." Pasteur's technique of weakening a strain by a damaging treatment or passing it through a susceptible host was termed attenuation and resulted in the creation of vaccines against anthrax, plague, yellow fever, rabies and many other diseases. Many vaccines have been developed over the years and children today receive a number of shots.

COMPOUNDS TO CONTROL MICROBES

By 1885, it was becoming clear that the causative agents of many illnesses were microorganisms. As scientists manipulated these microbes in the lab they found that certain dyes and other compounds were able to inhibit their growth. This inspired Paul Ehrlich to propose that chemicals may exist that will kill the microbe, but not the patient. The hope was that the chemicals will cure the illness of a patient. One of the diseases Ehrlich hoped to cure was syphilis, which had reached epidemic proportions in Europe. Little did he realise it would be a 17-year odyssey before he would develop salvarsan, the first effective chemotherapeutic agent. Salvarsan was the 606th chemical he tried.

This arsenic compound effectively kills *Treponema pallidum*, the causative agent of syphilis. The treatment, however, had many problems, causing long lasting health complications for those individuals who used it. In addition, despite *Treponema* being quite sensitive to salvarsan, the physician had to administer it intravenously for optimum effectiveness. Intravenous injection was a recent development and many doctors were leery of trying the procedure. In London, a young physician by the name of Alexander Fleming, then in the Army Medical Corps, was one of the few that was willing to treat patients. Fleming even got the nickname private 606 from his burgeoning practice.

His work validated the effectiveness of salvarsan against syphilis and convinced others to administer the treatment. Fleming but spent most of his time studying bacteria, and his success with salvarsan motivated him to search for other antibacterial agents. His first discovery was lysozyme, an enzyme produced by many organisms, including humans, that lysed some bacteria. This enzyme is not useful as a therapeutic agent because it is difficult to administer as a drug, but Fleming did develop titration methods and assays that would become very useful. Fleming's arguably most important contribution to science is his discovery of penicillin. In September of 1928, before leaving on a summer holiday, Fleming streaked some plates of *Staphylococcus aureus* and left them to incubate until his return.

In an improbable set of circumstances, the beginning of the holiday was cold, allowing some contaminating mold spores (that had blown in from a nearby window) to grow up on some of the plates. The temperature then increased encouraging the growth of the *Staphyloccoccus*. Many experimenters when confronted with a contaminated plate will look for the trash bin, but Fleming instead spent some time examining it. The fungus had a zone of clearing around it where the *Staphylococcus* colonies would not grow, suggesting the fungus was producing an antibacterial compound that had diffused into the medium.

Intrigued, he cultured the fungus, a *Penicillium* mold, and eventually isolated a soluble extract that could kill bacteria and treat localized infection. He called the new compound penicillin after the mold from which it came. Due to the technology then available, however, it was very difficult to prepare a solution that could be used throughout the body without causing problems.

World War II added a greater urgency to the search for compounds that could fight infectious disease. Wounded soldiers, if they survived the initial injury, would often develop life-threatening infections and there were no effective drugs to combat them. In 1939 Howard Florey and Ernst Chain began a systematic study of antimicrobial compounds in hopes of developing treatments for these soldiers and ran across Fleming's report written 9 years earlier.

They now were able to purify the compound completely and describe its high potency against microbes. The availability of penicillin during World War II saved countless lives. The rediscovery of penicillin touched off a search for other microbes producing substances that could kill or inhibit microbes, leading to the discovery of many more antimicrobials. With Florey and Chain, Fleming was awarded the Noble prize in Medicine and Physiology in 1945. In the ensuing decades the search and discovery of numerous antimicrobial compounds, combined with the development of vaccines, has eliminated many of the deadly diseases that plagued humankind. While we now realise this is a continuing war and not a one-time battle, our understanding of these microbes and the nature of disease will likely keep infectious disease at bay for the foreseeable future.

ENVIRONMENTAL MICROBIOLOGY

Early work in microbiology focused on pathogenic microbes and treatment of the diseases they caused. Martinus Beijerinck and Sergei Winogradsky began the transition from this early work into the present era where we study the molecular biology of a wide variety of microorganisms. Both were interested in microbes present in the soil and water and their work founded the discipline of environmental microbiology. Much of what we understand about bacteria in the environment and their impact can be traced back to the efforts of these two scientists. Martinus Beijerinck was originally trained as a botonist and began his work studying the microorganisms that were present in and around plants. He soon began experiments with microbes in the soil.

His greatest contribution was the development of enrichment media. Previously, microbes were cultivated on medium consisting of potatoes or extracts of left-over animal renderings. Such media would support the growth of many different bacteria, with chance and population density dictating what became dominant in the culture. In many cases this was exactly what was desired, but in other cases it was of interest to find bacteria capable of performing certain chemical conversions. Beijerinck discovered that by adding or removing certain compounds from the medium or incubating under different conditions, it was possible to favour the growth of certain microbes and prevent the growth of others. An example is Beijerinck's work with nitrogen-fixing bacteria, which are important in agriculture and the global cycling of nitrogen. They are capable of taking nitrogen gas (N_2) from the air and reducing it to ammonia (NH_3).

This is an important property since most organisms, including agriculturally important plants, can only use reduced nitrogen compounds such as ammonia. Beijerinck wanted to isolate an organism capable of fixing nitrogen in the presence of air, because all previous isolates fixed nitrogen only under anaerobic conditions. By making medium that did not contain a source of fixed nitrogen and then incubating in the presence of air (containing N_2), he demanded that any microbe growing in the medium had to be able to derive its nitrogen by performing aerobic nitrogen fixation. Using this method he succeeded in isolating a new microbe (*Azotobacter chroococum*) with these capabilities.

Similar selective culture techniques (as he liked to call them) enabled his laboratory to isolate sulfur-reducing and sulfur-oxidizing bacteria, *Lactobacillus* species, green algae and many other microbes. Selective culture techniques have since been applied to many different groups of microorganisms, allowing them to easily be brought into pure culture. Sergei Winogradsky was also interested in soil bacteria, especially those involved in the cycling of nitrogen and sulfur compounds. He was one of the first to isolate microorganisms responsible for the conversion of these elements in the soil, obtaining pure cultures of bacteria capable of the conversion of ammonia to nitrate by microorganisms in the soil.

Winogradsky also studied the consumption of hydrogen sulfide gas by sulfur-oxidizing bacteria directly in their natural habitat. When working with *Beggiatoa* (one of these sulfur-oxidizing bacteria) he discovered it was obtaining its cell carbon from carbon dioxide. He used the term autotrophy to describe this property, a radical idea at the time, and one not readily accepted by other microbiologists.

This led Winogradsky to propose the concept of chemolithotrophy; bacteria capable of growth using purely inorganic sources of carbon and energy. He realized that these microbes were essential for the cycling of nutrients in the environment, and the living world was dependent upon their action, long before others accepted the idea. As a result of the work of Winogradsky and Beijerinck there was great enthusiasm for identifying and classifying the bacteria inhabiting our natural world.

For the first part of the 20th century many scientists isolated microbes and made proposals for their organization into genera and species. In 1909 Sigurd Orla-Jensen suggested a classification scheme based on the functions and abilities of the bacteria, such as growth on certain compounds or the production of specific by-products. The Society of American Bacteriologists, later to become the American Society of Microbiology, applied this technique to prepare a report on the classification of bacteria. This eventually evolved into Bergey's Manual of Determinative Bacteriology in 1923 and subsequent editions have become authoritative reference works. There was great optimism that as more bacteria were brought into pure culture a clear organization, based on the evolutionary history of microbes, would emerge from studies of their physiology.

However, by the 1940's it became clear that bacteria were unwilling to go along with this idea because classification based on one set of properties was often inconsistent with classification using a different set of properties. Scientists threw up their hands and gave up the idea of finding a system of classification based on growth and morphological characteristics. Several decades passed during which bacteria became tools for understanding life, but their place in evolution was of little interest.

Then in 1961 Brian McCarthy and E. T. Bolton developed methods for the comparison of genetic material between species and Linus Pauling and Emile Zuckerkandl formalized the idea of using the makeup of biological molecules, their sequence, as a way to determine phylogenetic relationships.

Later, Carl Woese began to use this insight, choosing the sequence of the 16S rRNA from the ribosomes of bacteria as basis for evolutionary comparisons. This analysis was a watershed event for the evolutionary classification of bacteria and in 1977 Woese used this technique to assert that known bacteria should be divided into two separate domains that we now call Bacteria and Archaea, radically changing our view of the microbial world. Lynn Margulis proposed the remarkable notion that some bacteria fused with archaea to create eukaryotes.

This hypothesis, termed endosymbiosis, has been strongly supported by a range of subsequent biochemical data. In subsequent years, determination of 16S rRNA sequences, and then complete genomes, of microorganisms has enabled the natural classification of microorganisms. A review by Olsen, Woese and Overbeek provided a catalyzing summary of the tree of life for microbes, including eukaryotic microorganisms. This reorganization has reshaped our view of the tree of life for all organisms and helped to reassert the important role of microbes in shaping the past and affecting the future of the Earth.

MICROBES LIFE AT THE MOLECULAR LEVEL

In the early part of the 20th century techniques were developed to examine the inner workings of the cell and much of the work was performed in bacteria due to their experimental accessibility. Before this period, the method of turning genetic information into the proteins that carry out cellular processes was completely unknown. Indeed, the chemical composition of the genetic material was being hotly debated. From Gregor Mendel's work with pea plants, the nature of inheritance and heredity was understood.

However, it was unclear what molecule in the cell maintained and passed hereditary information on to subsequent generations. The leading contender for this role was protein, when in 1928 Fred Griffith discovered transformation in bacteria. Griffith knew that when *Streptococcus pneumoniae* strains are injected into mice, they cause rapid deterioration and death. Griffith isolated strains of the bacteria that were no longer capable of producing an outer slime layer and appeared rough when grown on solid medium in contrast to the smooth colonies of the original isolate.

When these rough strains were injected into mice, no illness resulted. Similarly, when a smooth microbe was heat-killed and then injected, the mice showed no signs of infection. A surprise was waiting for Griffith when he injected into mice a mixture of dead cells of a smooth isolate with live cells of a rough isolate: they died. When bacteria were isolated from these dead mice, they formed smooth colonies. Griffith hypothesized that the ability to create the slime layer was passed from the dead smooth cells to the viable rough cells, making them pathogenic again. The process became known as transformation. Griffith was ridiculed, as most scientists believed his preparation were contaminated with viable smooth cells.

It was not until 1944 that his student Oswald Avery and coworkers repeated the experiments of Griffith, reproducing his results and discovering that DNA was the material from the dead smooth cells that transformed the rough mutant. This was strong proof that the hereditary material in cells was DNA. Further experiments in the 1940's clearly established this observation with Joshua Lederberg discovering two more ways that DNA could be transferred between bacterial cells, conjugation and transduction. In 1943 Beadle and Tatum reported experiments with the fungus *Neuropsora crassa* that eventually established the idea that each gene in the DNA typically codes for one protein (the one gene-one enzyme hypothesis).

About 10 years later, Francis Crick, Rosalind Franklin, James Watson and Maurice Wilkins worked on experiments describing the structure of DNA and making predictions about how it was replicated. It was now clear that DNA stored the information for proteins and that proteins performed the many functions of the cell. The important question now became, how does one convert the information in DNA into protein.

A major contribution in understanding this puzzle was made by Paul Zemecnik and his lab who developed cell-free systems, first with rat liver and later using the bacterium *E. coli*. This allowed the study of translation in the test tube and he and other scientists used these systems to discover the important molecules involved in the process. An intense effort to describe these molecular events then ensued.

The first light came when Crick, Sydney Brenner and colleagues proposed the existence of transfer RNA (tRNA), a molecule that helps to create amino acid polymers based on nucleic acid sequence. Another critical player in the processing of information was revealed when Brenner, Francois Jacob and Matthew Meselson discovered that translation of genetic material into protein takes place on the ribosome and that the molecule being translated at the ribosome is RNA, not DNA. The next mystery was solved by Marshal Niremberg and J. H. Matthaei when they developed methods to decipher the genetic code that dictates the correspondence of nucleic acids to amino acids.

After the stunningly productive decade of the 1960's, the nature of the framework for the conversion of the genetic information into proteins was now understood and the basic mechanism has since been shown to be conserved across all biology.

CONTROL OF THE GENOME

Pathway begins in 1885 with the seemingly unimportant isolation of a common intestinal microorganism by Theodor Escherich. The microbe is eventually renamed *Escherichia coli* in his honour. It could not have been understood at the time, but this bacterium would play a central role in our understanding of life. *E. coli* achieved this position because it has several properties that made it invaluable as an experimental system:

- *E. coli is easy to take care of.* It can grow on a wide variety of media, at temperatures ranging from 20-45 °C, in the presence or absence of air.
- *E. coli grows quickly.* Under normal laboratory conditions it will go through a complete life cycle, doubling its population about every 30 minutes. With this rapid growth rate it is possible to generate large populations in just a few hours. Most experiments can be set up one day and the results observed the next.
- It can survive storage for long periods of time

without losing viability using relatively simple preservation techniques.

- *It was available.* There is an element of arbitrariness that this microbe came to serve such a vital role in biology. Many other bacteria probably could have filled this position, but *E. coli* was the one chosen.

The first of these was the discovery of autonomously replicated pieces of DNA separate from the bacterial chromosome. plasmids, as Joshua Lederberg called them, were capable of carrying genes that could be regulated and moved from one microbe to another independently of the bacterial chromosome. The ability of some plasmids to move between different cells meant that different bacteria could rapidly inherit genes carried on these plasmids. It was found that one class of plasmids encoded resistance genes that would enable the bacteria to grow in the presence of an antibiotic.

It later became clear it was possible to select for the presence of these plasmids by demanding growth of the bacterial strain on medium containing the antibiotic to which the plasmid encoded resistance. A second discovery that greatly impacted modern genetic engineering was that bacteria have their own parasites. As mentioned earlier, there are viruses that infect and kill bacteria, just as viruses infect animals and plants. The analysis of bacterial viruses was a major avenue to understanding the bacteria themselves. For example, investigation of the ability of some viruses to grow well on certain strains of bacteria and not others lead to the discovery of restriction-modification systems in bacteria. These systems allow bacteria to recognize and destroy foreign DNA and consist of two parts.

A modification enzyme that labels DNA as belonging to the bacterium and a restriction enzyme that recognizes a certain 4-8 base pair sequence in the DNA and cuts it in two if it has not been modified. The landmark paper of Hamilton Smith and Kent Wilcox showed that restriction enzymes were capable of cleaving double-stranded DNA into discrete pieces. The authors correctly hypothesized that the enzyme was recognizing a sequence of DNA and cutting it. Further studies on restriction enzymes led to the birth of genetic engineering. Joan Mertz and Ronald Davis made restriction enzymes into tools for the manipulation of DNA when they showed that many of the breaks these enzymes made in the DNA produced single-stranded complementaryends.

Stanley Cohen, Annie Chang, Robert Helling and Herbert Boyer then demonstrated that any DNA can be broken into fragments with restriction enzymes and by mixing that with a plasmid digested in a similar manner, it was possible to create recombinant molecules. Moving the plasmid into a microbe and growing it on selective medium could produce any desired amount of these recombinant molecules. Another discovery was "brewing" in the western United States. In the mid 1960s, Thomas Brock was on a trip to Yellowstone and became intrigued by the mats of green, brown and pink material that were found in many of the hot springs.

Brock was sure that these were living communities of microorganisms, yet the springs were at temperatures near boiling. Subsequent research proved his hypothesis correct and lead to the isolation of a microbe, *Thermus aquaticus,* capable of growth at 85 °C. Organisms capable of growing at this high a temperature were unheard of at the time and the discovery spawned two major developments. First, work with *T. aquaticus* and other unusual microbes by Woese and many others lead to the discovery of the Archaea. Second, the enzyme responsible for replicating DNA in *T. aquaticus* was found to be extremely heat stable, surviving temperatures of 95°C and copying DNA at 72°C.

The use of *T. aquaticus* polymerase sped the adoption of polymerase chain reaction (always referred to as PCR). This technique was developed in 1985 by Kary Mullis, and became incredibly useful for amplifying even small quantities of DNA. Originally the technique required fresh addition of polymerase enzyme after each step, which could become prohibitively expensive. The use of *T. aquaticus* polymerase made the procedure much more economical. PCR has become a pivotal technique in many areas of science, from detecting bacteria in the environment to the analysis of evidence at crime scenes.

In 1977 Walter Gilbert and Fred Sanger independently developed methods for determining the exact sequence of bases in DNA.

These techniques became immediately useful in determining the sequence of numerous important genes being investigated in countless laboratories. Further refinement has made DNA sequencing more efficient. It became efficient enough that in 1985 Robert Sinsheimer convened a meeting of a number of biologists active in genetics and gene mapping. Out of this meeting came a proposal to sequence the entire three billion base pairs of the human genome. This was equivalent at the time of proposing a mission to the moon, technically possible, but with unpredictable value. The initial cost of the project was estimated to be $10 a base pair or a staggering 30 billion dollars.

There was resistance from a significant portion of the scientific community, fearing that the project would siphon funds away from other worthy scientific projects. However, the clarity of the goal ignited the imagination of Congress and the public, creating unstoppable momentum for the project. After about a decade of effort, a preliminary draft of the entire genome was released in the spring of 2001, earlier than projected and dramatically under budget. The project itself drove the development of sequencing technology, allowing the determination of millions of base pairs of sequence per day at a cost of less than 3 cents per base at large sequencing facilities. Along with the human genome, many other model organisms have been or are about to be sequenced.

These include the genomes of many pathogenic bacteria, the mouse, the fruit fly and the nematode *Caenorhabditis elegans*, the last being an experimental model for development in eukaryotes. The analysis and application of this sequence data to investigate biological problems has resulted in the development of the field. Let's end this chapter with a defence of fundamental scientific research. That is, research without an obvious applied goal or application. If you look at the initial results generated in many of these important analyses, the research would have seemed esoteric and inconsequential. Is the analysis of microbes in the colon important? Who cares why a virus can grow in one strain of bacteria and not another? Why should the government support Tom Brock's isolation of microbes from the hot springs of Yellowstone National Park? What use is knowing the DNA sequence of a virus or a bacterial operon?

Yet, each of these insights began a thread of inquiry that changed the world. The take-home message is that basic research creates insights and applications beyond the imagination of even the scientists performing the work and should be supported. Microbiology is the study of organisms that at some point in their life exist as single cells and contain a nucleic acid genome that can replicate. Many organisms fall into this definition including algae, fungi, protozoa, bacteia and archaea. Together these organisms have a profound impact on the biosphere, making up the majority of life both in number and total mass.

Many illnesses are caused by infection with microbes and understanding these infections has lead to cures and better treatments. The emergence of new infectious agents will spur continued interest in microbiology. Many more microbes grow harmlessly in the environment, taking advantage of chemicals and/or sunlight to grow. Research into these microbes has also helped us understand the basic framework of life and revealed the basic fundamental rules that govern living systems. In the past microbes have been used in experiments to answer many scientific questions and they will continue to serve as excellent tools of inquiry in the future. A significant number of these discoveries have lead to important applications in many areas of human endeavor.

THE PERIPLASM

The periplasm is found in gram-negative bacteria and is the space in between the cytoplasmic and outer membranes. Many feel a periplasm is also present in gram-positive bacteria in between the cytoplasmic membrane and the peptidoglycan. This space is filled with water and proteins and is therefore somewhat reminiscent of the cytoplasm.

However, pools of small molecules in the periplasm are not like those in the cytoplasm because the membrane prevents their free exchange. Also, the proteins found in the periplasm are distinct from those in the cytoplasm and are specifically guided to this site during translation through. The peptidoglycan shell that provides the strength to prokaryotic membranes is also found in the periplasmic space of gram-negative bacteria, while in gram-positive bacteria it provides the outside border to the periplasm.

THE CELL WALL

This section will restrict itself to the bacterial cell wall, but at the end of the chapter we will compare this to archaeal cell walls. The cell wall is essential to the survival of most microorganisms. Many microbes live in environments in relatively dilute environments and the cell wall's most important function is to prevent the cell from bursting due to the osmotic stress placed upon it as discussed previously in section.

The cell wall also determines the shape of the cell. Any cell that has lost its cell wall, either artificially or naturally, becomes roughly spherical and will lyse due to osmotic pressure, unless placed in certain concentrated solutions. Finally, the cell wall helps to support any structure that penetrates from the cell out into the environment.

The structure and synthesis of prokaryotic cell walls is unique and many compounds found in the bacterial cell wall are found nowhere else in nature. It is true that plants also make cell walls, but they are chemically and structurally different. There are two basic types of bacterial cell wall structures that have been studied in detail: Gram positive and Gram negative. These two classes of bacterial cells look very different following staining with the Gram stain and this has been a standard basis for starting to identify different bacterial species.

When the Gram stain was developed by Hans Christian Gram in 1884 the molecular basis of the stain was unknown. In fact very little was understood about bacteria in general.

He just determined empirically that when bacterial smears were run through a four-step staining procedure using two different dyes, some cells retained the first dye and stained purple, while other only retained the second dye and stained pink. Years later it was discovered that the basis for this differential reaction relates to the cell wall.

The gram-negative cell has an additional layer and the outside of the cell appears convoluted when compared to the gram-positive cell. The gram-positive wall is much thicker than is the gram-negative wall and its external appearance is smoother. Gram-positive and negative cells do share one thing in common that is unique to bacteria - peptidoglycan. We will talk about the structure of this and then move on to examine the various structures found in each cell wall type.

Peptidoglycan is a thick rigid layer composed of an overlapping lattice of two sugars, N-acetyl glucosamine (NAG) and N-acetyl muramic acid (NAM), that are cross-linked by amino acid bridges. The exact molecular makeup of these cross-bridges is species-specific. NAM is only found in the cell walls of bacteria and nowhere else. Attached to NAM is a side chain generally composed of four amino acids. In the best-studied bacterial cell walls (*E. coli*) the cross-bridge is most commonly composed of L-alanine, D-alanine, D-glutamic acid and diaminopimelic acid (DPA).

The general peptidoglycan monomer showing the two sugars that make up the backbone. The R group will consist of 4 amino acids, with the best studied cell walls containing L-alanine, D-alanine, D-glutamic acid and diaminopimelic acid.

Note that peptidoglycan contains D-amino acids, which are different than the L-amino acids found in proteins. D-amino acids have the identical composition as L-amino acids, but are their mirror images. The use of D-amino acids is unusual in biology and bacteria have enzymes called racemases to convert between D and L forms specifically for this use.

The NAM, NAG and amino acid side chain form a single peptidoglycan unit that can link with other units via covalent bonds to form a repeating polymer. The polymer is further strengthened by covalent bonds between cross-bridges and the degree of cross-linking determines the degree of rigidity. In the *E. coli*, the penultimate D-alanine of one unit is linked to DPA of the next cross-bridge. In some gram-positive microbes there is a peptide composed of various amino acids that serves as a link between the cross-bridges. For example, in *Staphylococcus aureus* species five glycines make up the linker between peptidoglycan monomers. The sequence of these linkers varies considerably between species. The completed peptidoglycan layer forms a strong mesh that can be thought of as a chain link fence. The complete cell wall will contain one to many layers of peptidoglycan one atop the other, providing much of the strength of the cell wall.

While both gram-negative and gram-positive bacteria have peptidoglycan, its physical arrangement in the cell wall is different. In gram-positive cells the peptidoglycan is a heavily cross-linked woven structure that encircles the cell in many layers. It is very thick with peptidoglycan accounting for 50% of weight of cell and 90% of the weight of the cell wall. Electron micrographs show the peptidoglycan to be 20-80 nm thick. In gram-negative bacteria the peptidoglycan is much thinner with only 15-20% of the cell wall being peptidoglycan and it is only intermittently cross-linked. In both cases peptidoglycan is not a barrier to solutes, as the openings in the mesh are large enough for most molecules including proteins to pass through.

There are numerous antibacterial agents that target the bacterial cell wall because mammals do not synthesize walls and therefore are not susceptible to the toxic effects of these agents. Penicillin inhibits the linking of the amino acid side chains of peptidoglycan units, which therefore weakens the stability of the wall eventually, causing the cells to rupture. Humans and other animals even synthesize an enzyme that specifically attacks bacterial cell walls. The enzyme lysozyme is found in many body fluids and hydrolyzes the NAM-NAG bond in the cell wall. It serves as a critical part of the mammalian defence against bacterial invasion.

THE GRAM-POSITIVE CELL WALL

Another structure in the gram-positive cell wall is teichoic acid. It is a phosphodiester polymer of glycerol or ribitol joined by phosphate groups. Amino acids such as D-alanine are attached. Teichoic acid is covalently linked to muramic acid and stitches various layers of the peptidoglycan mesh together. Teichoic acid stabilizes the cell wall and makes it stronger.

Teichoic acid is a long, thin molecule that weaves through the peptidoglycan

Gram-negative Cell Structure

Gram-negative cell walls have a more complicated structure than do those of gram-positive organisms. Outside the cytoplasmic membrane is the periplasm, which contains the thin layer of peptidoglycan. The peptidoglycan in gram-negative cells contains less cross-linking than in gram-positive cells with no peptide linker. Covalently bound to the peptidoglycan is Braun's lipoprotein, which has a hydrophobic anchor in the outer membrane that helps to strongly bind the peptidoglycan to the outer membrane.

The cell wall in gram-negative bacteria contains much less peptidoglycan and is surrounded by an outer membrane. There is much less crosslinking between the peptidoglycan. LPS is also present in the outer membrane and penetrates into the surrounding environment.

THE OUTER MEMBRANE

The outer membrane is another lipid bilayer similar to the cytoplasmic membrane, and contains lipids, proteins, and also lipopolysaccharides (LPS). The membrane has distinctive sides, with the side that faces the outside containing all the LPS. LPS is composed of two parts: Lipid A and the polysaccharide chain that reaches out into the environment. Attached to Lipid A is a conserved core polysaccharide that contains KDO, heptose, glucose and glucosamine sugars. The rest of the polysaccharide consists of repeating sugar units and this is called the O-antigen. The O-antigen varies among bacterial species and even among various isolates of the same species. Many bacterial pathogens vary the make-up of the O-antigen in an effort to avoid recognition by the host's immune system.

LPS confers a negative charge and also repels hydrophobic compounds including certain drugs and disinfectants that would otherwise kill the cell. Some gram-negative species live in the gut of mammals and LPS repels fat-solubilizing molecules such as bile that the gal bladder secretes. This repulsion enables these bacteria to survive in this environment. The O-antigen and other molecules on the outer membrane are also used by certain viruses that infect bacteria, as a means to identify the correct hosts for infection.

LPS is medically important because when released from bacterial cells free LPS is toxic to mammals and is therefore called endotoxin. It creates a wide spectrum of physiological reactions including the induction of a fever (endotoxins are said to be pyrogenic), changes in white blood cell counts, leakage from blood vessels, tumor necrosis and lowered blood pressure leading to vascular collapse and eventually shock. At high enough concentrations the LPS endotoxin is lethal. Finally, the outer membrane keeps the enzymes in the periplasm from floating away from the cell.

There are fewer total proteins and fewer unique types of proteins in the outer membrane than in the cytoplasmic membrane. Porins are particularly important because of their role in the permeability of the outer membrane to small molecules. Porins are proteins that form pores in the outer membrane wide enough to allow passage of most small hydrophilic molecules All known porins have a similar structure, with the protein containing a central channel that allows the passage of molecules. This allows migration of these molecules into the periplasmic space for possible transport across the cytoplasmic membrane.

Some porins in the outer membrane are general, doing simple discrimination on size and charge, but having little substrate specificity. Examples include OmpF that is selective for positively charged molecules and PhoE that is permeable to negatively charged molecules. Other porins are more specific. The best studied is LamB, which recognizes the sugar polymer maltoolilgosaccharide and transports it through the outer membrane. Very large or hydrophobic molecules cannot penetrate the outer membrane, so the outer membrane serves as a permeability barrier to at least some molecules.

There are also other types of outer membrane proteins that are involved in various functions. OmpA in *E. coli* seems to connect the outer membrane to the peptidoglycan. Some pathogens contain outer membrane proteins that help them neutralize host defenses. Finally all gram-negative bacteria contain high molecular weight proteins involved in the uptake of large substrates such as iron-complexes and vitamin B12.

The differences between the cell walls of gram-positive and gram-negative bacteria greatly influence the success of the microbes in their environments. The thick cell wall of gram-positive cells allows them to do better in dry conditions because it reduces water loss. The outer membrane and its LPS helps gram-negative cells excel in the intestines and other host environments.

SOME BACTERIA LACK CELL WALLS

For most bacterial cells, the cell wall is critical to cell survival, yet there are some bacteria that do not have cell walls. *Mycoplasma* species are one widespread example and some can be intracellular pathogens that grow inside their hosts. Cell walls are unnecessary here because the cells only live in the controlled osmotic environment of other cells. Some of this obligate dependency is no doubt because of the lack of cell walls. It is likely they had the ability to form a cell wall at some point in the past, but as their lifestyle became one of existence inside other cells they lost the ability to form walls.

Consistent with this very limited lifestyle within other cells, these microbes also have very small genomes. They have no need for the genes for all sorts of biosynthetic enzymes, as they can steal the final components of these pathways from the host. Similarly, they have no need for genes encoding many different pathways for various carbon, nitrogen and energy sources, since their intracellular environment is completely predictable. Because of the absence of cell walls, *Mycoplasma* are quickly killed if placed in an environment with very high or very low salt concentrations and this feature also means that they have a spherical shape.

However, *Mycoplasma* do have unusually tough membranes that are more resistant to rupture than other bacteria since this cellular membrane has to contend with the host cell factors. The presence of sterols in the membrane contributes to their durability by helping to increase the forces that hold the membrane together.

Other bacterial species can occasionally mutate or change because of extreme nutritional conditions to form cells lacking walls, termed L-forms. This phenomenon is observed in both gram-positive and gram-negative species. L-forms have a varied shape and are sensitive to osmotic shock.

The Cell Surface Extends into the Environment

Surface structures are typically attached a membrane, and extend into the environment. Important structures include flagella, pili, fimbriae, and glycocalyx. These protrusions and surfaces interact with the environment around the microorganism and therefore are pivotal in how the microbe sees the world and how we see the microbe.

Flagella are one type of Structure used for Motility

Surface structures are typically attached a membrane, and extend into the environment. Important structures include flagella, pili, fimbriae, and glycocalyx. These protrusions and surfaces interact with the environment around the microorganism and therefore are pivotal in how the microbe sees the world and how we see the microbe.

Flagella are responsible for motility in liquid for many bacteria. There is a loose correlation between cell shape and the presence of flagella. Almost all spirilla, half of all rod-shaped bacteria, and only a few of the cocci are motile by flagella. In fact, most cocci are non-motile. One rationale for this correlation might be that spherical cells, such as the cocci possess, simply do not have the best geometry for directional movement by flagella, where more linear bacteria do. Flagella can be thought of as little semi-rigid propellers that are free at one end and attached to a cell at the other. The diameter of a flagellum is thin, 20 nm, and long with some having a length 2-3 times (about 10 μm) the length of the cell. Due to their small diameter, flagella cannot be seen in the light microscope unless a special stain is applied. Bacteria can have one or more flagella arranged in clumps or spread over the cell surface.

Chemical Structure

Flagella are mostly composed of the protein flagellin, which is bound in long chains and wraps around itself in a left-handed helix. The number of protein monomers that it takes to make a single turn of the helix is determined by the protein subunits themselves.

The long tail of the flagellum is attached to the cell through complex protein structures termed the hook and the basal body. One ring in the basal body rotates relative to the other causing the flagella to turn. The energy to drive the basal body is obtained from the proton motive force. In some fashion the translocation of protons from outside to inside the membrane causes the rotation of the flagellum.

One obvious model is that the protons move through the wheel-like structure of the basal body (similar to a water wheel, except using protons) and this causes the rotation of the assembly including the flagellum. When *E. coli* is swimming through a solution the flagella turn counter-clockwise and push the microbe through solution. This behaviour is termed smooth swimming. It is possible for *E. coli* to also reverse the direction of flagellar rotation and when the flagella turn clockwise, they pull against the bacterial cell. Since *E. coli* is flagellated peritrichously (that is, at many positions), it is pulled in all directions and tumbles.

How fast do bacterial cells move? They average 50 μm/sec, which is about 0.00015 kilometers/hr. This may seems slow but remember their tiny size. Also remember that this motility happens in water, which is much more viscous than air.

If a flagellum is cut off it will regenerate until it reaches a maximum length. As this occurs the growth is not from base, but from tip. The filament is hollow and subunits travel through the filament and self-assemble at the end.

Advantages of Motility

Typically microbes in aqueous environments will continually move around looking for nutrients. Even microorganisms in the soil will have uses and opportunities for movement. Sometimes this movement is random, but in other cases it is directed toward or away from something. As a rough guide, bacteria will want to move towards food or energy sources and away from toxic compounds. In other words, bacteria are capable of showing simple behaviour that depends upon various stimuli.

Chapter 2

Microbial Cell Structure

The origin of our present knowledge of the structure of cell membranes. Overton in 1895, was originally responsible for the idea that lipids were important in cell membrane structure. He made this suggestion because of the relative ease with which lipid soluble substances penetrated the cells. At that time there was no clear idea as to what types of lipid molecules were present or as to how they were arranged.

In 1917, Langmuir and other workers made deductions about the arrangement of molecules in molecular films spread on water surfaces and these have been described as being of fundamental importance to the modern concepts of the cell membrane structure.

Langmuir suggested that lipids at an interface of air and water arrange themselves in a monolayer with the polar ends of the molecules directed toward the water interface and the nonpolar ends at the air interface. In 1925, Gorter and Grendell found that the total amount of lipid they could extract from a red cell.

Ghost when spread in a monolayer occupied about twice the area of the membranes of the red cells. They suggested, therefore, that the lipid was arranged in the membrane as a bimolecular leaflet in which the hydrophilic polar groups were at the inner and outer surfaces and the hydrophobic carbon chains were directed toward each other in the interior of the membrane.

This was a simple and attractive model. It is difficult to imagine however how such a simple structure could account for all the various functions of the cell membrane.

Dr. G. D. Robertson has described the next stages of the development of our knowledge of the cell membrane. Briefly these are as follows: Harvey and Shapiro in the 1930's and also Cole during the same period studied the surface tension of starfish eggs and came to the conclusion that, at the ranges of pH's which apparently occurred inside cells, most pure lipoid substances gave surface tension values of the order of 5 dynes/cm^2. or more. But intracellular oil drops were found to have a surface tension of only 0.2 dyne/cm^2. The surface tension of whole starfish eggs as estimated by the force required to compress them, is even lower than this.

Thus, if only a purely lipid/water interface is concerned, these low values for the living cell cannot be explained. Danielli and Harvey demonstrated that there was some substance on the cell membrane which was responsible for such low values. This substance was a surface active agent of a cytoplasmic nature and appeared to be protein. Apparently in natural membranes all lipid polar substances were covered with at least a monolayer of protein. Danielli and Davson eventually proposed that cell membranes in general had the same sort of structure and that this consisted of one or more bimolecular leaflets of lipid with each polar surface having on it a monolayer of protein.

In 1936, Schmitt, using polarization microscopy, concluded that at least in red cell membranes the lipid molecules which were present had their carbon chains radially oriented. Thus we can picture the cell membrane as being composed of two layers of lipid molecules radially arranged (at right angles to the surface of the membrane) and a monolayer of protein applied to both the inner and outer surfaces of the membrane, with the long axes of the molecules lying parallel to the surface.

It had been suggested in the past that the cell membrane contained either pure lipid or pure protein, with pores of dimensions about the size of a large molecule, or that it was in the form of a mosaic containing either pure lipid or pure protein in various areas. However, this seems unlikely now except in some special cases. It is of interest that a common structure for the cell membrane exists in cells as diverse as erythrocytes, axons of nerve cells, muscle fibers, leucocytes of the blood, yeast cells, algal cells, the cells of higher plants and the ova of echinoderms, e.g., sea urchins and starfish. It seems very likely, however, that there are variations to some extent between the membranes of the various cells; nevertheless it is highly probable that a general structural pattern does exist for all of them.

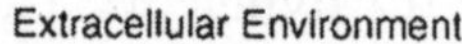

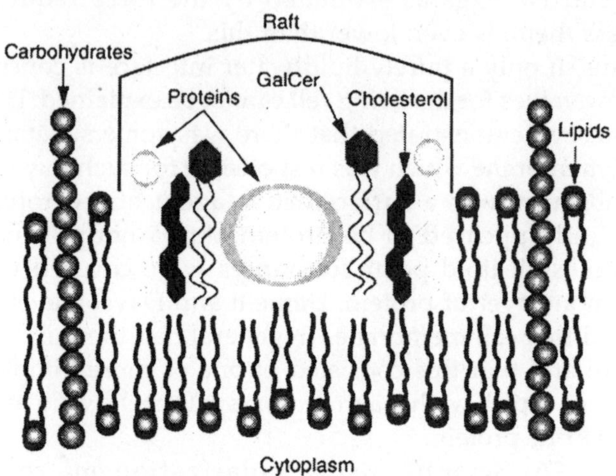

Fig. The Structure of the Cell Membrane

A typical example of variation was demonstrated by Mudd and Mudd who showed that erythrocytes are preferentially wetted by oil and leucocytes by water; this indicates that molecules at the surface of the membranes differ.

Studies with the electron-microscope have demonstrated that those membranes which have been studied have a thickness of the order of about 80 A (10,000 A = lã). Dr. Keith Porter has pointed out that the first electron-microscope photographs demonstrated the cell membrane as a solid structure, but that, as the technique improved and better resolution was obtained, the membrane was seen to be composed of two thin, very dense lines with a space between them. The lines measured about 25 A each and the space between them measured about 30 A across. It is tempting to suggest that the two outer lines represent the protein parts of the membrane and the central space the phospholipid portion, but this has yet to be proven. This electron-microscope pattern of structure and size is general in cell membranes–it is present, for example, in the nuclear membrane, the membranes of the endoplasmic reticulum and those of the mitochondria. In due course no doubt the electron-microscope will shed further light on the nature of the membranes and possibly the differences between the many types of membranes in various cells, but there are limitations to what it can do at the present moment.

Some of the ideas of the nature of the cell membrane have been derived from a study of its fundamental properties. First of all it is known that lipid-soluble substances have a preferential permeability across cell membranes and this suggests that there must be in the membrane a continuous layer which is composed of phosphatides, steroids or fats or combinations of those substances. Also cell membranes have a high electrical resistance which, Danielli points out, is further evidence that there must be a continuous layer of lipids. Another property is the existence of a low surface tension at the surface of the cell membranes.

We have mentioned earlier how Danielli has suggested that this property indicates that on the surfaces of the membrane protein layers are adsorbed. Cell membranes are disrupted by digitonin which has a special ability to form a complex with cholesterol and this suggests that this latter compound may play a part in the structure of the membrane.

Analysis of red cell ghosts has in fact demonstrated that a relatively large amount of cholesterol is present in the membrane. An interesting theory concerning the structure of the red cell membrane was proposed by Winkler and Bungenburg de Jong in 1941.

This theory gives the free cholesterol the role of stabilizer for the charged phospholipid molecules thus enabling them to give form to the membrane. This concept has been elaborated by Frey-Wyssling in 1953, whose studies on submicroscopic morphology and its relation to the chemical constituents of the cell are of fundamental importance. It is of interest to notice that in the two issues where cholesterol has been assigned the major role of structural component, it is present mainly if not entirely in the free unesterified form. There is one striking difference however in the nature of the cholesterol in the myelin sheath and the red cell.

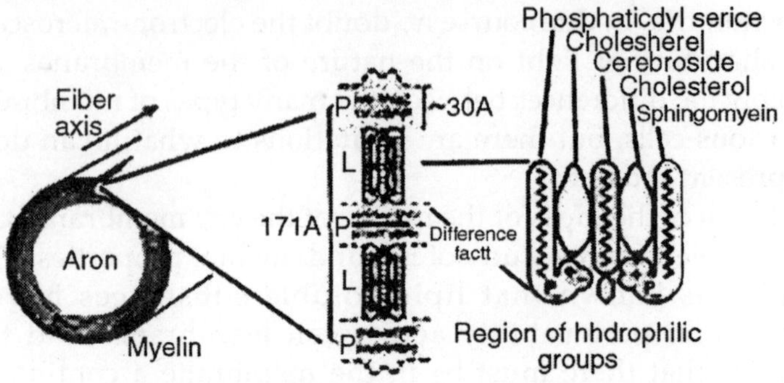

Fig. Molecular Structure of the Myelin Sheath

In the myelin sheath it is apparently stable and once deposited under normal conditions remains static for life. In the red blood cell there is a steady turnover and exchange with plasma cholesterol. In general, in tissues where there is active metabolism of cholesterol, esterified cholesterol is found, the amount varying from tissue to tissue.

One should, perhaps, at this point say a word about the composition of the myelin sheath since this is in a sense an extended membrane. Finean, from his studies on the x-ray diffraction pattern of myelin, suggested that it was probably composed of alternate layers of lipid and nonlipid, the nonlipid probably being protein. The lipid molecules appear to be curled or tilted so that their full length is not extended and there is probably a stable complex formed between the cholesterol molecule and the larger phospholipid molecules.

It is postulated that the free hydroxyl group of the cholesterol is important in binding the molecule by associating with the polar end of the lipid chain which it curls round. The hydrocarbon part of the cholesterol molecule is bound to the phospholipid by Van der Waal's forces. Cholesterol has a high dielectric constant and because of this it might act as an insulating agent.

THE PERMEABILITY OF CELL MEMBRANES

Danielli has discussed on several occasions the various problems associated with the permeability of natural membranes and the account which follows is taken largely from his writings.

PENETRATION OF COMPOUNDS

Danielli points out that there are three sites of resistance to free diffusion in cell membranes:

- The membrane/water interface for diffusion into the membrane;
- The membrane/water interface for diffusion from the membrane into the water;
- The interior of the membrane.

Every molecule which requires to pass into the interior of the cell has to get through these three sites of resistance. A molecule which has polar groups (-OH groups), as opposed, for instance, to nonpolar groups such as methylene, forms at least one hydrogen bond with water for each polar group; all these hydrogen bonds must be fractured simultaneously if the molecule is to penetrate even into the lipoid membrane. Glycerol, having three OH groups, must acquire sufficient kinetic energy to break three hydrogen bonds simultaneously before it can penetrate into the membrane. This involves a large amount of energy, consequently resistance 1 (diffusion from water to membrane) is so high for glycerol that resistances 2 and 3 are reduced to insignificance.

On the other hand, a molecule such as methyl alcohol which has only one OH group and one CH_3 group penetrates easily into the membrane and can also pass easily out of the lipoid layer into water and presumably ethyl alcohol (C_2H_5OH) can function similarly. Thus we find that for such molecules the rate of diffusion across the interfaces is very large compared with rate of diffusion across the membrane. The interior of the membrane in the case of these compounds is the most important factor controlling penetration. Differing from these examples are molecules which are predominantly hydrocarbon in nature such as carotene ($C_{40}H_{56}$).

With a molecule such as this, resistance I is insignificant; the molecule has no polar groups and thus even resistance 3 is not enormous; however, resistance 2 is very large indeed, because the hydrocarbon groups are hydrophobic and a considerable amount of kinetic energy is required to transfer CH_2 groups from lipid into water.

When many such groups are present they must all be transferred simultaneously from the lipoid layer into water, since otherwise the molecule remains substantially part of the lipoid layer and cannot diffuse away into the aqueous phase.With these examples in mind, Danielli classified penetrating molecules into four groups.

- Molecules with few polar and few nonpolar groups –for these resistance 3 is most important and they can penetrate comparatively rapidly e.g., oxygen and methyl alcohol.
- Molecules having a predominantly polar character-for these resistance 1 is most important and penetration is slow, included are glycerols, sugar and glycogen.
- Molecules having few polar and many nonpolar groups–resistance 2 is most important, penetration is slow e.g., carotene, vitamin A and fat.
- Molecules having many polar and many nonpolar groups–resistances 1 anti 2 are both important and penetration is slow e.g., polyhydroxylic bile acids, the glucuronide of oestrin and proteins.

Danielli has pointed out that these principles of membrane permeability have a distinct physiological significance. For instance, oxygen is required in large amounts by the cell and carbon dioxide must be disposed of rapidly and both of these substances can penetrate the cell membrane rapidly. On the other hand, the first products of glucose utilization, for example, by muscle cells are glycerol derivatives, these are valuable and the cell membrane does not let them get through too easily and so they escape only slowly. During sustained work lactic acid is formed–this would be toxic if it accumulated. The cell membrane is relatively permeable to lactic acid which thus can escape into the blood and this permits violent exercise to be maintained for a much longer period than would be the case if the cell membrane was impermeable to lactic acid.

Amino acids penetrate all membranes moderately well but protein penetrates badly and, therefore, amino acids are stored as protein. Fatty acids will penetrate moderately well, but neutral fat very poorly, therefore fatty acids are stored as neutral fat. Glucose which can get in and out of cells fairly rapidly is polymerized in liver cells to form glycogen which passes the cell membrane only with great difficulty. Each of these three latter products, to which the membranes are impermeable, is the results of polymerization of simpler compounds.

It is of interest that, when compounds which are diffusing through a cell membrane from outside to inside are being incorporated into some compound within the cell, they appear to penetrate the membrane more easily. A typical example of this can be drawn from amino acids.

If they are being actively incorporated into proteins, the rate of diffusion through the membrane will be much more rapid than if they were not being incorporated. Presumably, if they are being built into proteins, they cease to accumulate and so cease to build up a concentration gradient against the passage of further amino acids. Detoxification mechanisms also take advantage of the principles of the cell membrane permeability.

For instance, toxic substances which might penetrate the cell membrane become conjugated with amino acids, sulfuric acid, or glucuronic acid. Thus toxic cell-penetrating substances such as bromobenzene or menthol are converted into new molecules which penetrate the cells with great difficulty and once in the bloodstream tend to be filtered off by the glomeruli of the kidneys and cannot be reabsorbed from the urine and returned to the bloodstream by the kidney tubules.

Penetration of Ions Through Cell Membranes

Penetration of ions through cell membranes is of special interest because for instance, Na^+ions are more concentrated outside the cell (in the body fluids) and K+ions are more concentrated inside the cell, yet we know that K ions will pass into the cell and Na ions will pass out. Both these passages thus take place against a concentration gradient and this can only occur by the application of energy. It is claimed that oxidative enzymes and phosphatases play a part in the movement of ions in and out of membranes; a process which is known as "active transport."

Substances which interfere with the activity of these enzymes have been shown in many cases to interfere with the movement in or out of the cell of both sodium and potassium. There are certain specialized cell membranes, e.g., the cell membrane of the intestinal cells, in which the distal edge of the cell is folded to produce a larger number of extremely small fingerlike processes called microvilli; these microvilli are covered with a typical cell membrane. There are a large number of dephosphorylating enzymes localized around these microvilli. Their precise function in this position is not known, but it is almost certain, however, that they play a part in active transport by utilizing the energy derived from the hydrolysis of high-energy phosphates.

It is obvious from what we have said, therefore, that the cell membrane, in general, is a very selective structure anti exercises a good deal of control over the cell by deciding What goes in or out and by enabling the cell to be isolated from its environment. In this way it permits labour to go on in the cell which may be quite different from that going on in a neighboring cell, or in the body fluids surrounding the cell. Here then is the first division of labour in the cell to be described by us.

However the story is not yet ended. In 1952-54, Danielli proposed the structure of the cell membrane which explained the way in which large protein molecules in the cell membrane could permit the passage of ions through the membrane. Danielli conceived of a long protein molecule extending right through the membrane and passing outside it. This end became attached to an ion and then by contraction pulled the latter through into the interior of the membrane.

According to Lundgard and Hodgkin, a cytochrome oxidase energy-producing system is involved in the penetration of ions. They claim that their ion moving system requires adenosine triphosphatase (ATPase), creatine phosphatase and cytochrome oxidase. Conway has produced a redoxpump hypothesis that attempts to explain the penetration of ions through cell membranes by a mechanism which involves the successive oxidation and reduction of a compound in the membranes; possibly a phospholipid provides the energy for the movement of ions in this way. Phosphatidic acid is now thought to be concerned with ion movements through membranes.

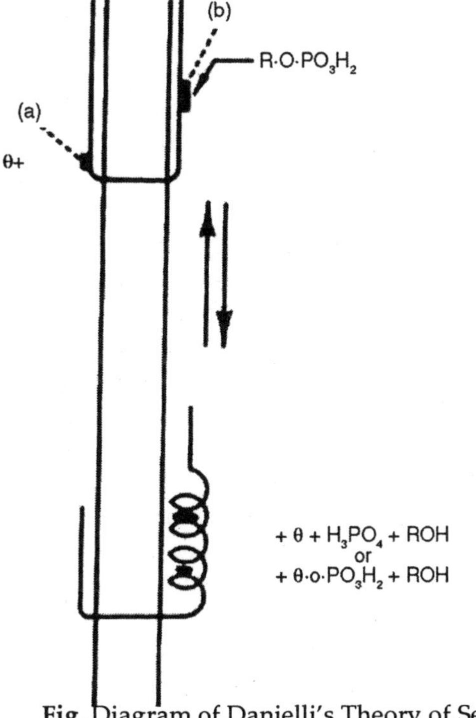

Fig. Diagram of Danielli's Theory of Secretion by a Contractile Protein

The possible relationship of phosphatases to cell membrane permeability has also been mentioned in connection with the microvilli (brush borders) of intestinal epithelium. The passage of glucose across the mebranes of gut cells could be explained by a mechanism involving these enzymes. It is of interest that in yeast cells too there is some evidence that phosphatases are localized on the membrane and that these enzymes are also involved in the permeability not only to glucose but also to phosphates. Phosphatases are also present on the brush borders of the kidney tubule cells where a good deal of absorption takes place.

More recent work has indicated that ribonuclease may play a part in the penetration of cell membranes by ions. Lansing and Rosenthal and Tenardo, for instance, showed that when ribonuclease was added from the outside to certain cells, it modified their permeability to ions; Brachet and Leduc found exactly the same thing for amphibian eggs. According to Brachet "amphibian eggs swell very much when they are immersed in a ribonuclease solution and since cell membranes often give strong cytochemical tests for ribonucleic acid it might well be that the integrity of this nucleic acid is of importance for normal permeability."

Danielli made a suggestion that the cell membrane consisted, as suggested before, of bimolecular leaflets of lipoid with protein molecules stretched both inside and outside the membrane, but that it also contains a series of pores which he calls polar pores. In other words, groups of protein molecules are oriented radially with their polar groups directed toward the interior of the pore and such molecules are thus able to control the penetration of the compounds or substances through this pore according to their polar group affinities.

Danielli discusses the association with all membranes of a group of enzymes, which are called "permeases." These compounds he says "permit either facilitated diffusion or active transfer of 'signal' molecules across the membranes." It is possible that insulin may function as a permease for cells of certain organs. To quote Danielli again "Once permease becomes incorporated in the plasma membrane the situation is transformed: specific substrates can penetrate, more permease will become available by induced synthesis, induced enzyme will appear and a whole range of further induced enzymes may appear in response to the action of the induced enzymes of the inducing substrate.

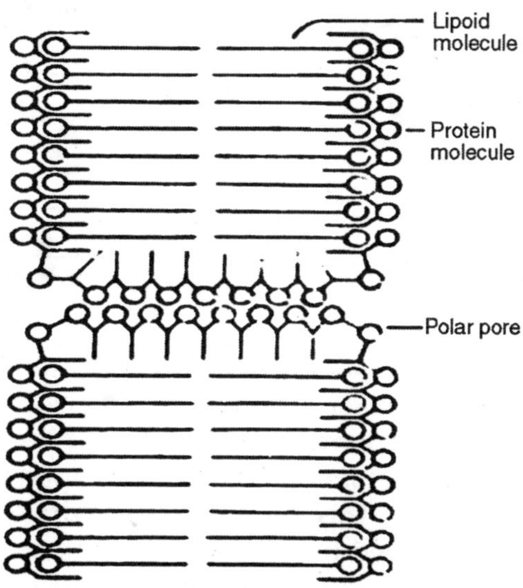

Fig. Diagram of Cell Membrane as Suggested

Danielli discusses the association with all membranes of a group of enzymes, which are called "permeases." These compounds he says "permit either facilitated diffusion or active transfer of 'signal' molecules across the membranes." It is possible that insulin may function as a permease for cells of certain organs. To quote Danielli again "Once permease becomes incorporated in the plasma membrane the situation is transformed: specific substrates can penetrate, more permease will become available by induced synthesis, induced enzyme will appear and a whole range of further induced enzymes may appear in response to the action of the induced enzymes of the inducing substrate.

Thus a transient infection with permease may potentially change drastically the state of differentiation of a cell–possibly even result in carcinogenesis." It is very likely that there will be considerable further developments in this subject of permeases and their relation to cell membrane permeability and these will be awaited with interest.

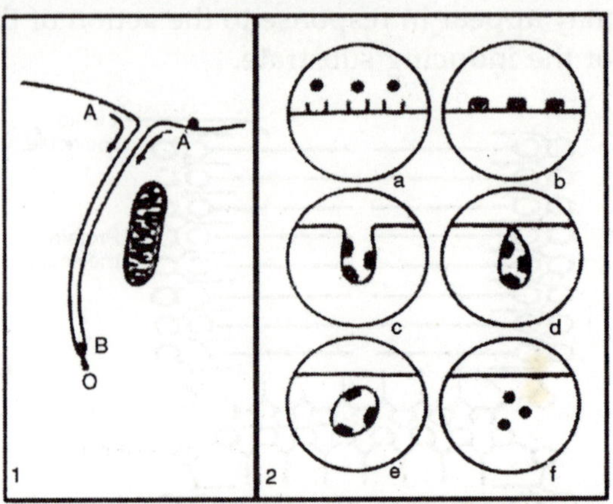

Fig. Diagrammatic Representation of Pinocytosis

A further development which should be noted when considering problems of penetration of compounds into cells is that of pinocytosis. This literally means "drinking by cells" and was first described by Warren Lewis many years ago in tissue culture cells. In this process, these cells were seen by a pseudopodial-like movement to engulf a droplet of culture fluid in which they were lying. The process was equivalent and similar to the phagocytosis of solid food by the *Amoeba,* only in the case of the tissue culture cell it is a droplet of fluid that is engulfed.

Now it seems that the cells of multicellular animals even in vivo can do the same sort of thing though there may be specialized parts of the cell membrane which carry out the process. A case in point is the kidney tubule cell. From the basal part of the cell, long internal extensions of the cell membrane pass a considerable distance up into the interior of the cytoplasm. They are present in some other cells too and are called "caveolae intracellulares."

The cell appears to be able to imbibe droplets of fluid by an engulfing movement of the membranes of these caveolae. It is possible too that the cell membranes of a variety of cells may be able to carry out pinocytosis even without the presence of these intracellular membranous extensions. Thus we see that, if a cell really "wants" to take in a few macromolecules that pass its membranes with difficulty, it can always take them by "the scruff of the neck" and pull them pinocytotically into its interior.

THE CYTOPLASM

Within the cell membrane and separated from the outside (extracellular) world by it, is the cytoplasm. Does it perform any labour? What is its contribution to the life and welfare of the cell? First of all let us find out what we can about its nature.

Cytoplasm chemically is composed of proteins, lipoids (which include fatty, phospholipid and steroidal compounds), carbohydrates, mineral salts and, of course, a good deal of water, (40-80%).

The constituents of the protoplasm of animals and plants are generally similar, although the relative proportions of the various components varies a good deal in different organisms and probably even in the same organism under different physiological conditions. Originally, according to Wilson in his classic work *"The Cell"*, the similarity between the protoplasm of plants and animals was stressed and particularly the importance of protein as a structural element in both of them (this, of course, can still be accepted).

However, the chemical evidence has since shown that some of the other elements differ in animals and plants, for instance, the plant protoplasm has rather more carbohydrate in it and proteins and lipids are the main constituents of the animal protoplasm.

Protoplasm, in general, behaves and appears to be something in the nature of a colloidal system which is very complex and behaves almost always as if it were a viscous liquid. This is particularly well demonstrated by living protoplasm undergoing streaming movements so well demonstrated in plant cells and in animal cells, such as Amoeba, when pseudopodia are being produced. Cells which are lying free tend to round up and become spherical when they are resting and if small fragments of protoplasm are chopped off from cells by a process described technically as "micrurgy," which simply means microsurgery, the little pieces so removed become spherical. Both these facts are in keeping with the conception that protoplasm is a viscous fluid.

Viscosity varies a great deal in different kinds of cells and it varies in the same cell from time to time according to the physiological state of the cell. Sometimes it may set into a semi-jellylike condition and this is particularly well shown by some types of slime molds (myxomycetes). When one of these is touched the whole organism suddenly sets into a gel formation and only after the passage of time does the organism gradually pass into a viscid fluid state again.

The ground substance of the protoplasm is known as hyaloplasm and contains a number of bodies and structures, vacuoles and so on and also a number of very tiny particles, some of them ranging into the ultramicroscopic level and which undergo active Brownian movement. According to Wilson "Flemming observed the dance of minute fat drops in living cartilage cells." Many of the theories of the nature of protoplasm developed in the latter part of the 19th century. At first there were not many theories about its structure because living cytoplasm is optically homogeneous or empty and between the years 1870 and 1890, when great interest was taken in examinations of fixed and stained tissues, a number of theories of the structure of protoplasm were developed based on its appearance after various types of fixation.

Probably one of the most wellknown of these is the fibrillar theory. This held that the protoplasm was made up fundamentally of delicate fibrils which were were either separate strands or formed a meshwork and this was situated within a homogeneous, optically empty, ground substance. Among the people whose names were associated with this theory were Leidig, Flemming, Carnoy and Heidenhain. These fibrils were thought to be of fundamental importance to the vital activities of the cell. However, later on, the fibrillar theory became subdivided into two subsidiary theories. One was the reticular theory and the other the filar theory.

The reticular theory was a modification of the filar theory in the sense that whereas the fibrils did not form networks in the filar theory, they did in the reticular theory. However, toward the end of the century a number of workers including Flemming, Bütschli and Fischer showed that coagulation artifacts could produce all the phenomena which had been described as fibrillar structures in the cytoplasm.

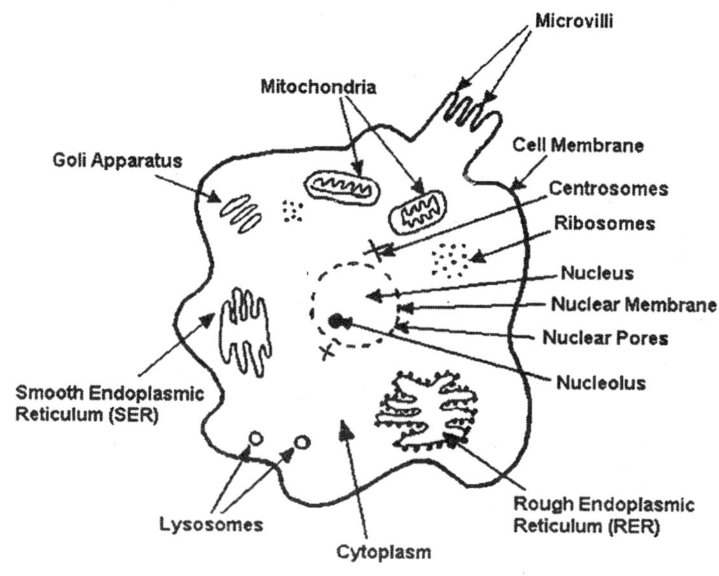

Fig. The Structure of Cytoplasm

A number of experiments had been carried out by various workers in which every conceivable type of fibrillar structure which had been described in the cell had been duplicated in non-living material made up, for instance, of white of egg or gelatine which had been coagulated by various fixatives. Thus by the end of the 19th century, the fibrillar theory and its offshoots were beginning to fall into disrepute.

Another theory of the structure of protoplasm was the alveolar or foam theory which was enunciated by Bütschli in a series of papers, the first of which he published in 1878. He believed that protoplasm was made up of a series of what he described as "alveolar spheres" which were suspended in a hyaloplasmic material. He regarded protoplasm as being equivalent to two viscid liquids, one of them forming the walls of the spheres scattered amongst the continuous substance (the hyaloplasm). There were also present in the cytoplasm a number of very small granules described as "microsomes," and we should note that the term "microsomes" as used by the early cytologists is quite different from the use of the term today. Nowadays "microsomes" have a fairly specific connotation since they are those bodies which are spun down in an ultracentrifuge from cell homogenates; the last of all the particles of the cell to be spun out. To what extent all modern microsomes represent something which really occurs in the living cell or not we shall discuss later on.

Another theory which developed in the 19th century was the "granule" theory. This was introduced by Altmann in three papers in 1886, 1890 and 1894 and was subsequently developed by quite a number of other writers, mainly Benda and Meves. Altmann, using a special technique, demonstrated red staining fuchsinophil granules in all the cells he examined. In many cases they were a fairly uniform size and in some cells they were very closely crowded and appeared to occupy nearly all the cytoplasm. Altmann regarded these granules as elementary organisms and he called them "cytoblasts" or "bioblasts" and believed that they "lived" in a homogeneous ground substance again called the "hyaloplasm."

We know now, of course, that Altmann's granules were actually mitochondria. Wilson in 1928 summing up the theories of cytoplasmic structure said "we are driven by a hundred reasons to conclude that the protoplasm has an organization that is perfectly definite but it is one that finds visible expression in a protean variety of structures. We are not in a position to regard any of these as universally diagnostic of the living substance." He went on to say that the fundamental structure of protoplasm lay beyond their limits of microscopic vision. His final conclusion was that protoplasm possessed an ultrastructural configuration known as a "metastructure."

The attempts to demonstrate definite structure in

protoplasm were followed by a reaction against the alveolar, fibrillar and granular theories and led back again to the extreme reverse conception that the material was structureless. Yet during this pre phase-contrast, pre electron-microscope period, some prophets were found amongst the biochemists who are, or at any rate were not noteworthy for having a morphological outlook on cells.

These prophets demanded the presence of a cytoskeleton which they felt was a necessary structure along which various enzymes, etc., should become aligned. One of the first steps which helped to bring back to the conception that structure existed in apparent hyaloplasmic cytoplasm was the development of the phase-contrast microscope in which a number of formed structures were seen to exist in the cytoplasm of the living cell.

Immediately beneath the cell membrane, the protoplasm of the cell is differentiated to form an "ectoplasm" which appears to be in a partly gelled condition. It is probable that this ectoplasm plays an important part in the movements of the cell and it appears to be actively forming and reforming in the movement of protocells such as the Amoeba. It is also modified to form cilia. These are, in effect, ectoplasmic prolongations surrounded by a cell membrane. The ectoplasm is also a constituent part of the microvilli which occur on absorbent cells of the kidney tubules and intestine.

Phase-contrast microscopy demonstrated a good deal of structure in the apparent hyaloplasm and later this was supported by the results obtained with a further development of this sort of microscopy–the interference microscope in which objects of different density in the cell were coloured differently according to their density. More recently the electron microscope, using the excellent fixation produced by osmium tetroxide has demonstrated an incredible complexity in what was only a few years ago thought of as a relatively homogeneous hyaloplasm.

The cytoplasm is filled with mitochondria, microbodies of various sorts, double membranes, Golgi apparatus and various granules; in fact, there is relatively little of what we might describe as hyaloplasm. This hyaloplasm does, however, exist and when fixed and examined under the electron microscope it seems to be composed of an extremely fine network and this becomes coarser or finer according to the type of fixative used.

Now what is this network composed of? Wykoff has pointed out that a number of proteins which contain long, flamentous type molecules are capable of forming gels which have many of the physicochemical properties characteristic of protoplasm. He points out that gelatin is one of these and describes one of his experiments in which gelatin gels were first fixed with osmic acid and then examined under the electron microscope.

He found that the gelatin was arranged in a network and that the pores of the net were small or large according to the concentration of the gelatin/gel or the fixative used. With a concentration of 2 to 4% in a gel, the net which results possesses a pore size which is very similar to that found in cells, but the pores get smaller if the gel is more concentrated. It is probable therefore that the network-like structure of the hyaloplasm which we see is really to some extent the effect of the fixative, although it is partly the expression of the nature and distribution of the filamentous molecules of protein which form the basis of the hyaloplasm.

These molecules are, in fact, probably arranged in a sort of network, a structure that Sir Rudolph Peters once described as the "cytoskeleton," and which he thought was the basis of cytoplasmic structure. This suggestion was made some years before the electron microscope demonstrated such a wealth of formed components within the hyaloplasm itself. Within the network of filamentous molecules (both protein and polysaccharide) which build up the hyaloplasm, a variety of molecules, not only free protein, polysaccharide and lipid but also ions, move about attaching and detaching themselves to the network in accordance with the electrical and other forces operating at that level.

We can, indeed, picture this hyaloplasmic network as a sort of skeleton for the rest of the cell. Now in this hyaloplasmic structure lie the rest of the cell elements which have to be separated from it to carry out the types of activity in which they specialize.

THE ENDOPLASMIC RETICULUM

A prominent feature of the cytoplasm of cells, particularly pancreatic cells, under the electron microscope, is a basophilic fibrillar structure which takes us back a little into the fibrillar theory of protoplasm; this structure is known either as the "endoplasmic reticulum" or "ergastoplasm" and reference will be made again to these names later on.

The structure and function of this material has been worked out and recognized in the last few years with the aid of the electron microscope, although it was originally discovered a long time ago with the light microscope. The discoverer was Garnier and the year 1897. The material discovered was a basophilic fibrillar material which could be seen in stained cells, particularly in the basal region of gland cells and it was called by Garnier "ergastoplasm."

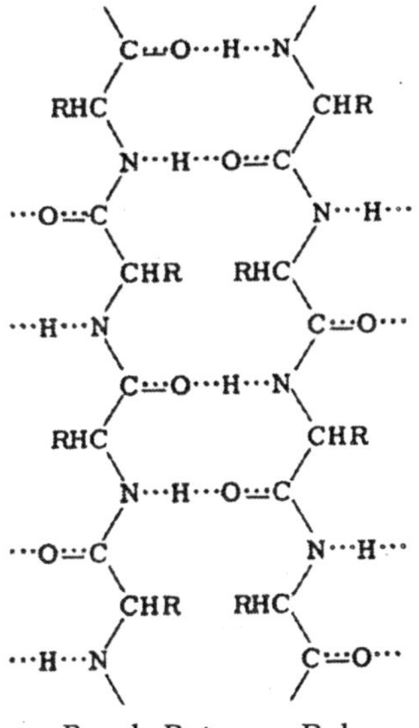

```
          \                  /
           C=O···H···N
          /              \
     RHC                    CHR
          \              /
           N···H···O=C
          /              \
 ···O=C                     N···H···
          \              /
           CHR    RHC
          /              \
 ···H···N                   C=O···
          \              /
           C=O···H···N
          /              \
     RHC                    CHR
          \              /
           N···H···O=C
          /              \
 ···O=C                     N···H···
          \              /
           CHR    RHC
          /              \
 ···H···N                   C=O···
          \              /
```

Fig. Hydrogen Bonds Between Polypeptide Chains

The ergastoplasm appears to be part of the system described as the endoplasmic reticulum. This latter structure was described by Porter and Thompson in a paper in 1947 on electron-microscope studies of a chick macrophage. The macrophage when very thinly spread out showed that the cytoplasm was everywhere permeated by a tenuous network which they described as a lacelike reticulum.

This they quite understandably called an endoplasmic reticulum. Later on, when fine sectioning of cells became possible in the 1950's, the endoplasmic reticulum seemed to occupy the position that was normally occupied by the "ergastoplasm," as Garnier had called it.

In addition it also appeared to spread through the other parts of the cytoplasm of the cell and to be present, to a greater or less extent, in practically all the cells which have been examined. However, at this point we should try to solve the problem of terminology.

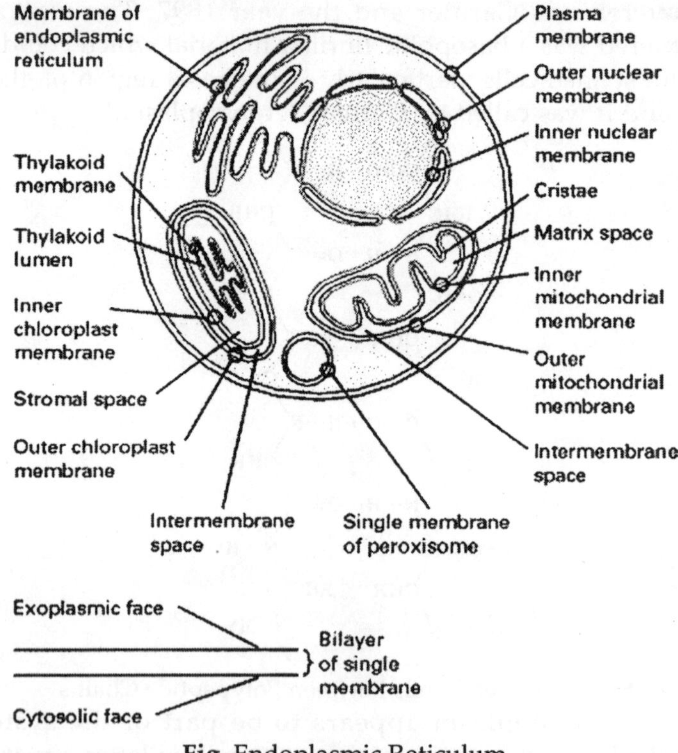

Fig. Endoplasmic Reticulum

One of the characteristics of ergastoplasm is that it is basophilic, but not all the endoplasmic reticulum is basophilic although both basophilic and nonbasophilic regions in ultrathin sections of cells have the same double membrane structure.

The difference is that the basophilic portions have ribonucleoprotein particles attached to the membranes and the nonbasophilic parts do not. Therefore the ergastoplasm can be thought of as a specialized basophilic part of the endoplasmic reticulum. Sjóstrand tried to avoid this controversy by giving the names of Greek letters to various membranes in the cells. The membranes which form the endoplasmic reticulum he has described as the á-cytomembrane, the membranes of the mitochondria and of the cytoplasm receiving other Greek letters to designate them. Sjóstrand's recommendation for nomenclature, however, seems to be a rather cumbersome method of describing these structures.

The use of the term "ergastoplasm" for the system of cytoplasmic membranes is largely supported by the French school recently headed by the late Charles Oberling, whereas in the United States the accent is more on endoplasmic reticulum with ergastoplasm as a specialized part of it. This is the nomenclature followed by the Rockefeller group headed by Porter and Palade.

Since the nature of the endoplasmic reticulum and, indeed, of many other cell structures is based on electronmicroscope studies, we might perhaps briefly consider the significance of electron-microscope pictures, in general and attempt to assess to what extent they represent a real structural condition in the living cell. The late Dr. Oberling has listed six points which suggest that the electron-microscope picture is a true picture of the living cell.

- The types of structures which are seen under the electron microscope have also been seen in living cells and they have been photographed or filmed by the phase-contrast microscope.
- Whenever the possibility has existed for comparison

of the same cells, for instance, alive as viewed under the phase-contrast microscope or in fixed condition with the electron microscope, there has been perfect aggreement in the picture.

- The same methods of fixation and observation reveal all the structures side by side and present in all cells, even in those which are very different from both the phylogenetic and the functional points of view. The appearance of these structures depends to a large extent on the perfect preservation of cells by first class fixation.
- In order to attain excellent pictures the cells must be fixed in the living state. This proves the great sensitivity of the observed structures and the reliability of the procedures in detecting such structural changes which take place immediately after vital functions have ceased.
- The techniques of homogenization, fractionation and ultracentrifugation have given the opportunity to isolate these structures and obtain them in a relatively pure condition in sufficient amounts to permit biochemical investigations, thus bringing closer the collaboration between the morphological and biochemical studies of the cell.
- These structures, as they appear under the electron microscope, do not always produce the same aspect, but they vary according to the evolutionary phases and also differ in pathological conditions from which the cell may have been suffering.

These points are very strong ones and possibly the strongest of all from the point of view of the very delicate structure which is demonstrated in electron microscopy is point 4. With regard to the first three, very few people have ever doubted that, for instance, mitochondria (as seen with the light microscope) existed in living cells. They have been seen in living as well as in fixed cells examined by the optical microscope and it was no surprise to find that they were also present in electron-microscope pictures. The point that has been really at issue is whether the extremely fine structures which the electron microscopists demonstrate do really exist in life. The very fine, double membrane structures and so on, could possibly be produced as a result of the technique used, although in view of the widespread constancy of findings this is unlikely. Nevertheless we should still keep a healthy attitude of caution in accepting all these fine structural details and be prepared to alter our views should evidence accumulate that there are errors in this depiction of ultrafine structure. At the moment, however, it appears that the electron microscope is telling the truth so far as we can interpret it, but that it is not yet showing everything that is in the cell.

Garnier, the discoverer of ergastoplasm, brought out some very important points concerning its nature and these have been listed in an excellent article by Haguenau to which the reader is referred for further details of the ergastoplasm. Garnier, according to Haguenau, said that, in the basal region of all cells, without exception, there was a portion of the cytoplasm which seemed to have a fibrillar or rodlike structure, that these filaments or rods were not separate structures but were actually part of the cytoplasm and were in direct continuity with it.

Haguenau points out that the electron microscope has provided an excellent confirmation of this. Garnier also said that these filaments were stained by basic dyes–those used were safranin gentian violet and toluidine blue–and that the intensity with which this basic staining occurred varied with the stage of secretion in gland cells.

Another point which he made was that the ergastoplasm was not a permanent structure and that its development was related directly to the state of activity of the cell. He said, for instance, that in gland cells the filaments appeared much more numerous when the cell, having already gone through a cycle of secretion and excretion, was preparing a new cycle. When the cell became loaded again with secretory granules, the filaments became less obvious and disappeared.

His final point was that the filaments were closely related from a topographical point of view with the nucleus, that masses of the material formed laterally on the sides of the nucleus and that sometimes the latter was completely encircled by the ergastoplasm. Garnier also thought that nuclear sap or chromatic substance originating from the nucleolus was able to pass through the nuclear membrane and enter into association with the ergastoplasm. This was actually a very significant comment as will be seen later.

Following Garnier's work many other authors figured and described and categorized the ergastoplasm. Prenant in 1898 wrote a review on the ergastoplasm which he called the "protoplasme supérieur." He described the ergastoplasm as a very important zone of the cytoplasm which was capable of differentiating into specific structures and among these, according to Haguenau, were included the "Nebenkern," the "Dotterkern" of the germ cell and the ergastoplasm of the gland cell and even the Nissl bodies of nerve cells.

These were rather interesting conclusions since they have now been supported by electron-microscope studies. Haguenau has pointed out that in the history of most discoveries there is a period in which it is first described, then a lot of other people describe it and finally there is a period when everybody believes it is an artifact due to fixation. She points out that at about the same time as the ergastoplasm was discovered Altmann discovered mitochondria, but comments made on the latter by Benda in 1898 and 1899 served to complicate the proper interpretation and acceptance of the ergastoplasm.

Haguenau divided the post-Garnier workers into three groups: the first group were represented by Morelle in 1927 who claimed that the ergastoplasm was nothing more than modified ground cytoplasm which took up basic stains, this being due not to any difference in structure but simply to a chemical difference. He claimed that this area of the cytoplasm was never really fibrillar. Then there was a second group who thought that the ergastoplasm was only mitochondria which were distorted in shape. Champy in 1911, for example, stated that mitochondria and ergastoplasm were one and the same and that the preparations which showed ergastoplasm were simply preparations with poorer fixation than those which showed mitochondria.

To some extent there is some justification for this point of view, since mitochondria and ergastoplasm are so closely related to each other topographically that it is natural enough for a mitochondrial stain to demonstrate concentrations of mitochondria in the same site as the ergastoplasm. The third group agreed with Garnier that mitochondria and ergastoplasm were quite different structures but these, according to Haguenau had to fight very hard to prove it. Prominent amongst these was Regaud.

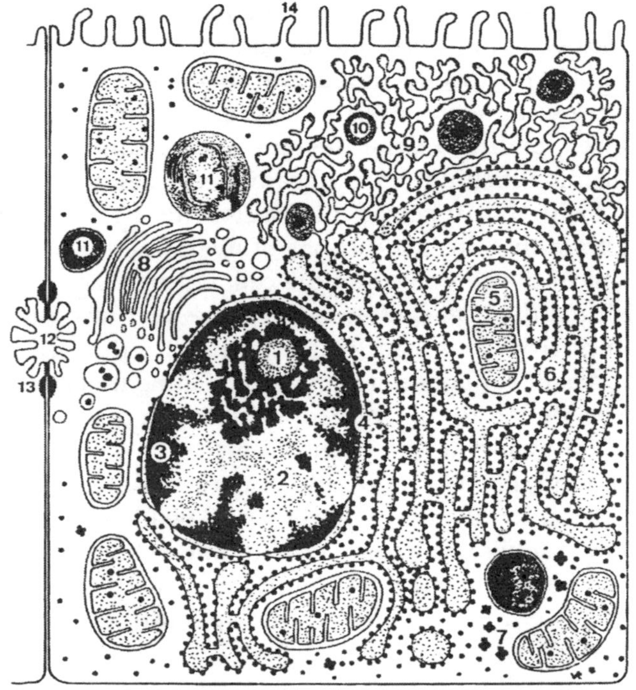

Fig. Endoplasmic Reticulum of Liver Cell

1. Nucleolus; 2. Chromatin; 3. Dense Chromatin; 4. Nuclear Pores; 5. Mitochondria; 6. Rough Endopla-smic Reticulum; 7. Ribosomes; 8. Golgi Apparatus; 9. Smooth Endoplasmic Reticulum; 10. Peroxisomes; 11. Lysosomes; 12. Bile Capillary; 13. Desmosomes; 14. Microvilli.

The critical experiment which Regaud published in 1908 was one in which he showed that, if he had acetic acid in his fixative, there were no mitochondria in the preparation but the ergastoplasm appeared quite normal. On the other hand, with acetic acid absent mitochondria could be demonstrated very conveniently but the ergastoplasm did not take up its normal appearance.Haguenau quotes Regaud, "It is possible that the ergastoplasm consists of a protoplasmic support impregnated with chromatin or a closely related substance."

Rees in 1940 using the polarizing microscope made a study of the ergastoplasm and came to the conclusion that there was a homogeneous filamentous structure present in the cytoplasm of living cells, but it was the electron microscope which finally confirmed and helped to delineate the structure of the ergastoplasm. Preliminary studies by Porter and colleagues leading to the designation of this material as endoplasmic reticulum have already been mentioned.

In 1950, Hillier was probably the first person to demonstrate that, in fine sections of liver, fine fibrous matter is present in the cytoplasm of the cell, but he was not able to establish its identity. Dalton in the same year also produced electron micrographs which demonstrated quite clearly that the cytoplasm contained a number of filamentous units and that these tended to be grouped in particular areas and were reduced following fasting of the animal.

The work of the French and American schools, together with contributions by the Swedish school headed by Sjóstrand, have now established a great deal of interesting information about the structure, nature, identity and distribution of the endoplasmic reticulum and its specialized part, the ergastoplasm.

The endoplasmic reticulum appears fibrous in nature in sections and it is of interest that these fibers appear to form pairs and they have been regarded, in fact, as paired membranes. Actually, it seems that they are in many cases the walls of membraneous tubules or even sac-like structures or vesicles and are frequently referred to as double membranes. It is of interest that these membranes show the 75-80-A unit structure which is so characteristic of cell membranes and which may be significant in the light of something we will say later on about them.

There is some discrepancy between the findings of electron microscopists regarding the ergastoplasm and the findings of the classical cytologists. The latter claim that the basophilic staining with which these fibrillar structures are associated appears to be localized at the basal end of the cell, whereas with the electron microscope these fibers or membranes extend all through the cell.

This, however, can be explained in part by the fact that it is only when there are quite a number of membranes close together that they show up as a strongly basophilic area under the light microscope and in other parts of the cell the membranes are more widely spread out and are not organized in dense groups; therefore the characteristic staining reaction is spread out over a great area of the cytoplasm and becomes diluted.

Pancreatic Acinar Cell Model:

Ca^{2+} Regulated Signal Transduction Pathways

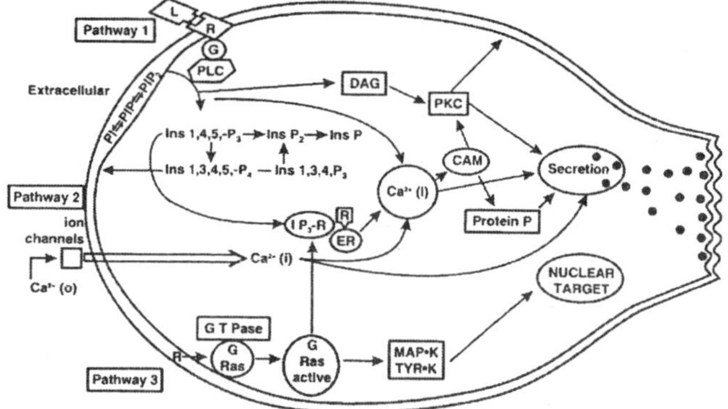

Fig. Portion Pancreatic Acinar Cell

Furthermore some of the membranes do not have ribonucleic acid granules attached to them (the component which produces the basophilic staining) and are called "smooth" endoplasmic reticulum. It is also evident that some endoplasmic reticulum membranes may show as basophilic structures but do not have any obvious RNA granules –so, perhaps, this material can be associated with the endoplasmic reticulum in a molecular as well as a particulate form.

In areas where the ergastoplasmic membranes are concentrated together so that the whole cytoplasm has a lamellar appearance, it has been suggested that the term "organized ergastoplasm" should be used. This term was suggested by Houdson and Ham in 1955.

Another interesting organization of the ergastoplasm is the "Nebenkern." Nebenkern are structures which are basophilic in nature. Their significance and method of formation was unknown to older cytologists, but under the electron microscope it appears that they are concentric layers of ergastoplasmic membranes which in section look like an onion bulb or, as some authors put it, a finger-print.

It is possible that these structures may originate from mitochondria. The ergastoplasm may also be modified in nerve cells in the region where Nissl bodies are found, as demonstrated by Palay and Palade.

There are considerable variations in amount of ergastoplasm which is thought to be related to the differentiation of cells and it has been suggested that it may also be linked to growth. There seems little doubt that the membranes are linked to function and they are particularly obvious in cells which are engaged in the production of protein.

Noteworthy also about the structure of ergastoplasm is the association with it of a large number of granules. It is of interest that these granules are associated with the outer part of the membrane. The side of the membrane directed toward the cavity which the membrane lines in the cytoplasm is smooth and has no granules attached to it, whereas on the cytoplasmic side of the membrane, quite distinct granules are found. They are basophilic in nature, range from approximately 120 to 150 A in diameter and have been the subject of considerable speculation.

Since they are basophilic in nature, it is obvious that they are, as mentioned earlier, the cause of the basophilic staining of the ergastoplasm recorded by the older cytologists. The studies by Palade and Siekewitz over the last five or six years have demonstrated the very interesting fact that these granules are composed of ribonucleic acid combined with protein. The way in which Palade and Siekewitz obtained this information is of interest.

They treated homogenates of cells with deoxycholate, a surface-tension reducing agent which detached the granules from the membranes. Then they were able, by differential centrifugation, to isolate a number of them, carry out chemical analyses and so establish that they were largely composed of ribonucleic acid. However, one should not think that all the ribonucleic acid in the cytoplasm is associated with the endoplasmic reticulum. It is present in other parts of the cell as well, but there is some evidence that about 25% of the cytoplasmic ribonucleic acid is, in fact, associated with the endoplasmic reticulum.

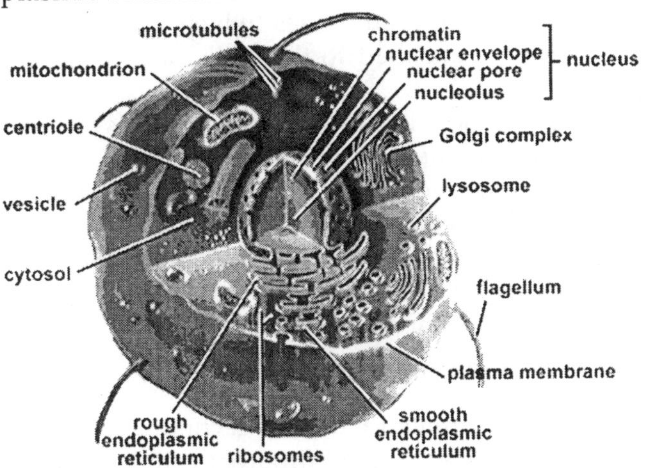

Fig. Hypothetic Cell

One of the interesting observations about these particles, demonstrating perhaps not so much division of labour in cells as co-operation of labour between cell constituents, is that in the nucleolus, ribonucleic acid particles are found which are the same size (150 A in diameter as those associated with the endoplasmic reticulum.

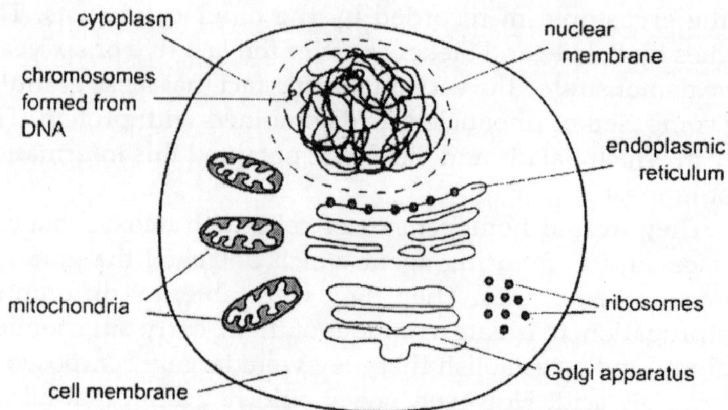

Fig. Principal Organelles in a Cell

(It is of interest, also, that the deoxyribonucleic acid particles which are found in the chromatin of the nucleus are also about 150 A in diameter.) It seems possible that the RNA particles could pass out from the nucleolus through pores present in the nuclear membranes (which we will talk about later) and become attached to the walls of the endoplasmic reticulum.

The fact that there is a coincidence in size is of very great interest from this point of view. A different type of migration has also been suggested by Gay who showed with the electron microscope that blebs forming at the surface of nuclei may become converted into endoplasmic reticulum and that RNA granules appeared to become attached to them before they actually became detached from the nucleus.

THE NATURE OF MICROSOMES

When Claude applied Bensley's technique of differential centrifugation to cells, isolated different cellular components and subjected them to chemical analysis, he described a series of minute bodies which are submicroscopic in size, i.e., below the limit of the resolution by the optical microscope. He described these bodies as microsomes. They are the smallest bodies in cellular homogenates and come down after everything has been centrifuged off, i.e., at the highest speed of centrifugation, leaving only the supernatant.

Many chemical studies were done on the microsomes and a good deal of information on their enzymes and chemical composition was obtained. However, one thing which made cytologists and particularly electron microscopists uneasy was the fact that, if one looked at a liver cell (which Claude used largely as the source of material for the production of microsomes) under the electron microscope, nothing could be seen in the cytoplasm which bore any resemblance to the microsomes. Since these were of a size which could be seen at the magnification used by the electron microscope, it was difficult to understand where they came from in the homogenates. The microsomes contain a good deal of ribonucleic acid and so does the ergastoplasm.

Finally, by isolating the pellet of microsomes centrifuged out from the homogenized cells and by examining ultrathin sections of them under the electron microscope, various workers showed that the microsomes were small vesicles with ribonucleic acid granules attached to them. It thus seemed quite obvious that the grinding-up of the cell during homogenization broke up the endoplasmic reticulum and the bits formed themselves into little vesicles which floated in the fluid and were finally centrifuged down. This work was first done by Slatterbach in 1953 and subsequently by Palade and his colleagues.

Relation of Endoplasmic Reticulum Membranes to Cell Membrane

The E.R. (endoplasmic reticulum) membranes have a structure which is similar in structure and dimensions to the cell membrane. It has been suggested that the E.R. membranes might represent complex infoldings of the cell membrane and that the cavities between the folds of the membranes may be in direct continuity with the exterior of the cell. We know that some infoldings occur, they have been mentioned before and are known as the caveolae intracellulares.

These structures have a space of constant size between the two folds of cell membrane which form their walls–this space is 200 A in width, thus the total width of the system including the two membranes of 80 A thickness each is about 360 A. Now, in some parts of the cell the membranes and spaces of the endoplasmic reticulum approach this dimension, but there is tremendous variation in the space (known as the "cisternal space") between the membranes, in parts it may be greatly dilated.

This suggests that the endoplasmic reticulum is not necessarily part of the same system to which the caveolae intracellulares belong. No certain connection between the E.R. membrane systems and the cell membrane has been established although the possibility exists and Porter and other authors have recognized that the connection may not be permanent but that it may be spasmodic.

A distinct possibility occurs also that the endoplasmic reticulum developed originally from infoldings of the cell membrane, that it then increased in complexity and finally lost its connection with its point or points of origin, but temporary direct connection between the cisternae of the endoplasmic reticulum and the exterior of the cell can occur.

It is thus possible that the cavity between the membranes in the E.R. represents a path by means of which the contents of the membranes or sacs (cisternae) can be excreted to the exterior. In other words the secretory products of the cell pass from the cytoplasm through the walls possibly in molecular form, collect into droplets in the interior of the membranes and pass along them. It is, of course, possible that secretion products form in the cisternae.

In 1957, Hendler and colleagues found that, in the gland cells which produce albumen in the hen oviduct, the cavities or spaces between the membranes were dilated and contained a precipitate which possessed the staining and other characters of the secreted material which is found in the lumen of the gland. There is other evidence to indicate too that, possibly, these spaces between the membranes represent a pathway for secretory material.

It is of interest, as shown by Brandes that, in the ventral lobe of the prostate gland in mice and rats, the spaces between the pairs of membranes are so enormously dilated that the membranes from opposite pairs come into close apposition with each other; pairs of membranes still exist but they are the halves of opposite pairs which have come together. The space between the original pairs of membranes is packed with an amorphous material. The ribonucleic acid granules in the such pairs of membranes are, of course, on the inside instead of the outside as they were when the original halves of the pairs were together. It is possible that this enormous accumulation of material in these endoplasmic sacs really represent accumulation of secretion.

It is of interest that, if the animals are castrated, the pairs of membranes with the granules on the inside separate from each other, the amount of secretory material appears to decrease and the original halves of the membrane pairs come close together again so that the ribonucleic acid granules can now be seen on the outside. Brandes has also demonstrated that, by injection or implantation of male sex hormone, the original state of the cell could be produced again.

This is in a sense an " Alice Through the Looking-Glass" type of cell in which the normal arrangement of cytoplasm and endoplasmic reticulum spaces is reversed. When this cell is in its normal condition, the spaces between the membranes contain, in fact, the strands of cytoplasm and by far the main body of the cell is occupied by what are presumably the secretion products.

It has probably not been made clear up to date that all these endoplasmic sacs are probably continuous with each other and thus form a convoluted system of spaces coursing through the cytoplasm of the cell. It appears, also, that folds of the endoplasmic reticulum constitute the nuclear membrane (this will be discussed later). Since this membrane has been shown to contain pores, it is quite possible that material could pass from the nucleus into the cisternae and along these channels to the exterior of the cell without having to traverse the cytoplasm or the cell membrane at all.

At this point we should enlarge a little on the subject of the caveolae intracellulares. These were originally described by Yamada as consisting of a number of infoldings of the surface membrane of some cells. He found them in endoneurial, endothelial, pulmonary epithelial and muscle cells. Although it is possible that these caveolae could, in fact, represent the regions where the E.R. channels communicate with the exterior, the existing evidence is against this. If this were so then there is a continuous, if extremely tortuous, pathway which stems from the vicinity of the nuclear membrane extending to the exterior of the cell and which could be described as a possible circulatory system for the cell.

At the moment it seems most likely that the caveolae represent only blind pockets which extend for variable distances into the cells and there is no evidence that they are actually continuous with the ergastoplasm. Palade has suggested that these invaginations in endothelial cells, at any rate, represent stages in a process of what is described as "micropinocytosis" which has already been discussed. It has been suggested that when these little vesicles containing water pass into the cell, the cell membrane then dissolves away and the water is liberated into the cytoplasm.

This conception of the cell membrane "pinching off" and a little vesicle passing into the cytoplasm is attractive, but one of the difficulties of accepting it is the mechanism of the formation of the vesicle. We know that the cell membrane can invaginate in this way (it has been seen to do so) but whether, when the membranes on the two sides of the vesicle bend around and come in contact with each other, they are, in fact, capable of coalescing and nipping off a vesicle is another matter. It has been pointed out that, if the membrane of a cell was a purely lipid substance, there would be no difficulty at all in such a concept, but, since the outside and probably the inside parts of the lipid membrane of a cell are covered with protein, this would make such a coalescence rather difficult to conceive from a physicochemical point of view.

On the other hand, an invagination of the membrane may draw off a little droplet of water and the membrane may then burst internally and squirt the drop of water into the cell and then there would be no difficulty in the burst sides of the membrane joining up in the usual way. This seems to the present author to be a much more likely way for pinocytosis to take place than for a vesicle to be actually cut off.

The Golgi apparatus has also been shown to be composed of double membrane structures without, however, the associated ribonucleoprotein granules of the endoplasmic reticulum. Although we will be dealing with the Golgi apparatus later on, we might mention here the concept which has been put forward by some authors that the Golgi apparatus represents a part of the system of the endoplasmic reticulum and that it is in communication with it.

We do not know whether the whole of the endoplasmic reticulum is in continuous communication with itself, however, if it is and if the Golgi apparatus is part of it, then it is possible to assume that everything passing along the cisternae of the reticulum will have to filter through the Golgi region. It should be stressed that there is no proof that this is so, that many cytologists believe it is not so, but that such a possibility exists.

One problem which should be discussed at this point is the origin of the endoplasmic reticulum. Where in fact does it come from? There are many controversial theories on this subject. It has been suggested that it might originate as an infolding or infoldings of the cell membrane and this theory was put forward by Palade.

The Nebenkern because of its high concentration of E.R. membranes might be considered its a possible region in the cytoplasm where this material is being formed. The Nebenkern is made up of a dense aggregation of concentric rings and one could, perhaps, consider the possibility of the membranes peeling off from such a germ centre. The same sort of appearance has also been noted in association with the nuclear membrane and has led to the suggestion that the endoplasmic reticulum is formed in this region and that the layers are, in fact, peeling off the nuclear membrane. However it may quite easily be interpreted the opposite way, i.e., the close apposition of the existing ergastoplasm around the nuclear membrane may be a part of the complex canalicular system which permits an almost direct passage through the nuclear membrane direct into the cisternae of the reticulum. Again, there are a number of authors who think the endoplasmic reticulum is associated with mitochondria.

There is no doubt that many pictures of mitochondria, closely surrounded by concentric layers of endoplasmic reticulum have been obtained and we, ourselves, have found this particularly well demonstrated around the mitochondria of the liver cells of scorbutic guinea pigs. In some cases, Sheridan in this department has found a tremendous concentration of many layers of reticulum around the mitochondria as though, in fact, the reticulum is being formed on the surface of the mitochondrial membrane and is being split off. Rouiller and his colleagues have demonstrated that, in animals which have been poisoned, the endoplasmic reticulum, destroyed or badly damaged by the treatment, always reappears in association with mitochondria.

The membranes of the reticulum may not undergo direct physical formation in the sense that the membranes are produced on the surface of the mitochondrial membrane and split off; it may be that the mitochondria supply the energy necessary for the production of these membranes. Furthermore the close relationship between the endoplasmic reticulum and the mitochondria may simply be physiological, that is they are co-operating in some metabolic process such as the synthesis of protein. However, the origin of the endoplasmic reticulum is a problem which is far from solved and we must await further evidence before drawing a conclusion.

THE MITOCHONDRIA

Mitochondria have been known for many years. They were first discovered and described by Altmann in 1886 and were put more or less definitely on the cytological map by Benda in 1903. They can be easily demonstrated with suitable dyes. Altmann's aniline fuchsine-picric acid technique shows them up very well. Regaud's iron-hematoxylin method is also very good. The mitochondria can even be seen in the living cell by staining them intravitally with Janus green and, of course, they show up extremely well with the phase-contrast and interference microscopes. There is not much difficulty, then, in establishing their existence and nature. Their dimensions vary, of course, but they range in most cells from about 0.5 to 2ì or longer.

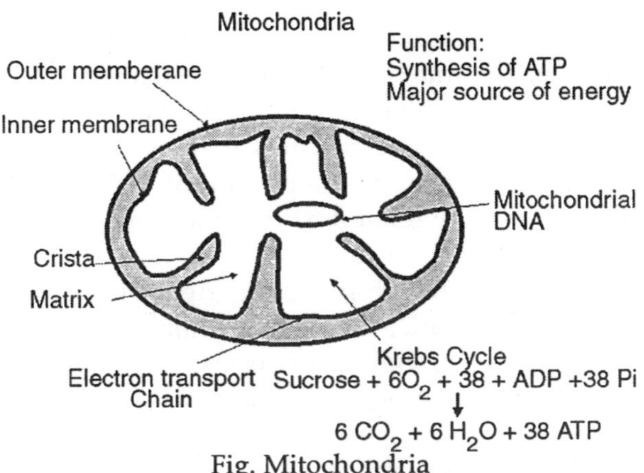

Fig. Mitochondria

The filamentous mitochondria, for example, which are present in connective tissue cells may be much longer than this. Extremely small and even submicroscopic mitochondria may also exist. Extremely smallsized mitochondria are, in fact, known and Rhodin has described structures ranging from 0.1 to 0.5ì which are presumably mitochondrial in nature. Green has pointed out that the various enzymes and proteins which mitochondria contain can be accommodated in a very much smaller particle than any known mitochondria and still perform all the functions required of them.

Some of these much smaller mitochondria may not have all the details of organization of larger mitochondria but there is no reason why they should not contain all the enzymes and other compounds which are important for mitochondrial structure and function.

It is of interest to note that, in living tissue culture cells, the mitochondria undergo continuous movement. This has been noted by quite a number of workers and the movements have been described frequently. These movements may be fitful individual movements by the mitochondrion itself, i.e., bending, wriggling and writhing, or may consist of movement of the whole mitochondrion through the cytoplasm of the cell. The cause of the bending and twisting movements of the mitochondria is not exactly known. At one time it was thought to result from the contracting of the polypeptide chains which are part of the membrane of the mitochondria. It may be due to the fact that the mitochondria were engaged in active ion pumping or being twisted by cytoplasmic movements. In the second type of movement in which there is a transport of the whole mitochondrion from one part of the cell to another, the movement is possibly related to streaming movements of the cytoplasm or to electrical forces.

The mitochondria have been described as making journeys from the cell membrane to the nuclear membrane and back again almost as if they were discharging either an electric charge or even perhaps some compound on the membrane that they touch.

There are also records that, when the mitochondria come in contact with the nuclear membrane the nucleolus sometimes moves across the cell and comes in contact with the nuclear membrane at the same point at which the mitochondrion is touching it and this suggests that there may be some exchange of compounds or substances during this period. This has been demonstrated in spinal ganglion cells by Dr. Tewari in the author's laboratory.

Filamentous mitochondria have frequently been observed to break up into batonettes and these have been observed to break up into granules. The reverse process has also been found to occur, granules have been seen to join up into batonettes and these into longer filamentous mitochondria. What the significance of this breaking up is we do not know, but a little later we will refer to this process again in the light of what will be said about the metabolic activities of mitochondria.

When mitochondria were studied in ultrathin sections of tissue, it was found by Palade in 1953, Sjöstrand, 1953 and by Rhodin in 1954, that they had an interesting internal structure. These workers found first of all that there was a single membrane round the outside, that inside this membrane was another membrane so that the two made a double membrane and, after a little controversy, it was agreed that the inner of these two membranes was extended to form a number of bars or plates which projected into the interior of the mitochondria, in some cases touching or almost touching the other side. These plates or bars or tubes had double structure again, since they were a reflection of the inner membrane of the mitochondria.

They were described by Palade as the "cristae mitochondriales". Changes in these cristae appear to be related to function, for instance Palade has drawn attention to the fact that the amount of cytochrome in the mitochondrial fraction of a homogenate is directly related to the number of cristae present in the mitochondria.

When metabolic activity is high, for example, in rapidly contracting skeletal and heart muscle, the mitochondria have many densely packed cristae. In smooth muscle, however, where the activity is greatly reduced, the cristae present in the mitochondria are relatively sparse. There are claims that, e.g., in mouse tumor cells and in paramecium, cristae may be everted into the cytoplasm. Presumably in this case the outer membrane is folded into cristae.

It is of interest that the double membranes which surround the mitochondria show the same 80 A unit structure which is characteristic of the cell membrane and Robertson has suggested that it is possible that the mitochondria may have originated by an invagination of the membrane which was nipped off to form the mitochondrion.

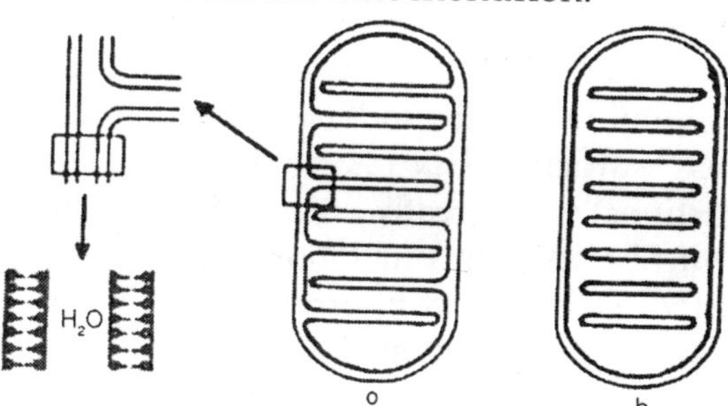

Fig. Structure of Mitochondria

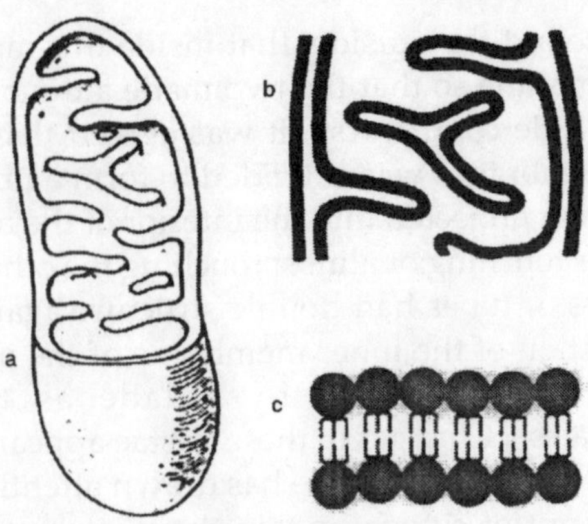

Fig. Three Dimensional View of Structure of Mitochondria which Appears as a Bag filled with Fluid.

This is an interesting suggestion but it is not readily acceptable to most cytologists and is, in fact, considered unlikely by many of them. The membranes of the mitochondria appear to consist of two protein layers separated by a double layer of lipids. This is again characteristic of cell membrane structure. Bensley in 1934 was the first one to isolate mitochondria by differential centrifugation and he was able to demonstrate that mitochondria contained a considerable amount of protein, lipid and fat. Bensley's chemical composition of mitochondria is

Proteins and unknowns	65%
Glycerides	29%
Lecithin and cephalin	4%
Cholesterol	2%

Subsequent studies have shown up to 30% of phospholipid (lecithin and cephalin). Mitochondria undergo a very considerable change under various conditions. Bensley claimed that mitochondria (as demonstrated by standard staining reactions) were greatly reduced in liver cells of starved animals, but Sheridan in our laboratory and Fawcett have demonstrated by electron-microscope techniques a great increase of mitochondria in starvation.

Sheridan has further demonstrated a still further increase of mitochondria in the liver cells of guinea pigs suffering from scurvy. Mitochondria are also affected in number and form by narcotics, enzyme poisons, old age and so on. Sometimes mitochondria store unusual compounds and sometimes they store compounds which are usual but to an abnormal degree. The double membrane enclosing the mitochondria is of a semipermeable nature and as a result of this mitochondria have been described as osmometers.

There is no doubt that molecule of various sizes pass with different degrees of readiness through the mitochondrial membrane. It has been shown that large molecules pass through fairly easily while some small molecules pass through either not at all or only with very great difficulty. It has even been said by Emmelot and Bos that this permeability, particularly of liver mitochondrial membranes, is affected by thyroxine. Also there are claims by some authors that mitochondrial membranes have pores. Mitochondria may accumulate water and swell. They may do this in starvation and in various pathological conditions. When this occurs they produce the condition which the pathologists describe as cloudy swelling. Cloudy swelling has been known to occur sometimes in neoplasms and sometimes after poisoning with various toxins.

The mechanics of cloudy swelling are that the mitochondria become very large and lose their cristae, the matrix stains less strongly and sometimes the outer membranes of the mitochondria may disappear and the mitochondria may fuse and so form structures known as chondriospheres. Under some pathological conditions mitochondria may become disrupted, undergo lysis and portions may be ejected from the cell. Mitochondria are capable of abnormal storage. Accumulation of ferritin by mitochondria was demonstrated by Kuff and Dalton in 1957 and they may also accumulate iron pigments or bile pigments.

Brachet has described ferritin particles in young erythroblasts where he says they penetrate the mitochondria and then break into smaller iron-containing particles. When the mitochondria eventually burst, the granules become scattered throughout the cytoplasm and take part in the formation of hemoglobin. Silver granules and keratinous materials have been also seen inside mitochondria. They have been found to contain melanin and Graffi in a series of papers from 1939 to 1941 claims that they also contain or store carcinogenetic hydrocarbons. Sometimes this storage affects the function of the mitochondria and sometimes the accumulation of material is due to malfunction of the organelle. They may also store neutral fats or lipids. In plants mitochondria have been recorded as storing starch.

As a result of storage mitochondria undergo physical changes. The membrane may become single and the cristae decrease in size and may be lost altogether. If the stored substance is eliminated, the mitochondria resume their normal appearance. There is a vast literature on the role of mitochondria in the production of almost any kind of product–protein, yolk, fat, glycogen and so on–but this is not the place to review this work.

There have been many scornful remarks in the literature about the supposed synthetic activity of mitochondria, but the study of the enzymic equipment of mitochondria shows that they have the equipment to effect a wide range of synthetic activities. It has been mentioned that the isolation of mitochondria by the differential centrifugation of homogenized cells was first carried out in 1935 by Bensley and by Hoerr. It was not until 10 or 15 years later that this technique became used extensively and this time by biochemists who set to work to find out the metabolic composition of mitochondria, i.e., the nature and content of enzymes concerned with metabolic processes.

THE CHEMICAL NATURE OF MITOCHONDRIA

In Bensley's work it was suspected that mitochondria contained a good deal of lipid material. One finding which suggested this was that mitochondria were stained by a method which is similar to that used for staining *Mycobacterium tuberculosis* and *Mycobacterium leprae*. These bacteria are stained by treating them for some minutes with a hot phenolic solution of basic fuchsine and once stained in this way they resist the de-staining effects of acid alcohol. This is the origin of their designation as acid-fast bacteria and their staining idiosyncrasies are believed to be due to a waxy or lipoidal coat. Mitochondria stained by hot acid fuchsine resist the decolouring action of picric acid for longer periods than most other cell constituents and this suggests that the mitochondria might also have a membrane which contains material of a lipoidal nature or at any rate which is of similar composition to that of the bacteria mentioned.

Mitochondria are difficult to demonstrate with conventional methods of staining if the fixative has contained acetic acid, alcohol, ether, chloroform, acetone, or other fat solvents. Also mitochondria have been found to stain with osmic acid and, incidentally, to resist de-staining by extraction with turpentine–a property of certain types of lipoprotein complexes.

Baker has found that his "acid-haematin" test for lipids applied to a variety of tissues, nearly always gave a positive, reaction with the mitochondria. Thus there had accumulated for some years direct and indirect evidence that mitochondria contained appreciable amounts of lipid. Bensley's experiments, which, incidentally, were carried out some years before Baker's staining studies, showed that mitochondria which had been produced by homogenization and differential centrifugation contained appreciable amounts of lipid (30%).

METABOLIC SUBSTANCES FOUND IN MITOCHONDRIA

Perhaps we should give a brief summary of what was suspected about the metabolic significance of these organelles prior to the revolutionary studies which were made with isolated mitochondria beginning during the middle of the 1940's. One of the characteristic staining reactions of these organelles is that they give a green or green-blue stain with Janus green B which is diethylsafranine azodimethylaniline and it has been claimed by Cowdry that this reaction is primarily due to the diethylsafranine monocarboxylic acid component since this compound gives a very good and specific reaction with these structures. It was observed many years ago by T. B. Roberston that if one drop of a saturated solution of safranine was added to a solution of trypsin, a coloured precipitate was formed and subsequently it was demonstrated that this coloured precipitate had proteolytic activity.

Then Marston demonstrated that other azo dyestuffs would react in a similar way, particularly neutral red which is a dimethyldiaminotoluazine hydrochloride. Marston suggested that these results indicated that the reaction of mitochondria with Janus green might signify that they contained proteolytic enzymes. Now, the concept of the staining reaction with Janus green has been pretty well proved by Lazarow and Cooperstein to indicate that mitochondria play an important part in cellular oxidations and that the production of a pink colour from the Janus green by the mitochondria is due to the DPN specific dehydrogenases.

The fact that Janus green B does not stain the mitochondria permanently green but that the green colour gradually becomes reduced to the pink and then to the colourless form has been known for many years. Joyet-Lavergne demonstrated more than twenty years ago that mitochondria contain an oxidase system which oxidizes cobaltous to cobaltic salts and that these latter stain the mitochondria green.

It was noted by Gatenby that in the snail, *Lymnaea*, the mitochondria are coloured yellow in the natural state presumably by a carotenoid pigment; also extracted lipomitochondria frequently have a yellow appearance which too is probably due to carotene. It is of interest in this connection that it has been demonstrated by the present author and by Joyet-Lavergne that mitochondria give a blue reaction with antimony trichloride in chloroform solution, an acknowledged reaction for vitamin.

A. Carotene is also a provitamin A, so these histochemical results indicate the presence of vitamin A in mitochondria. Criticism of Joyet-Lavergne's results was made by Gomori because the former author had used alcohol as a fixative. The present author has, however, always applied antimony trichloride in chloroform solution direct to fresh unfixed tissues and this suggests that this result is a true one, in any case biochemical tests have now confirmed that mitochondria (prepared by homogenization and differential centrifugation of the cells) contain 27-32% of their weight of lipids and that 100 mg. of this lipid contains approximately 249-910 U.S.P. units of vitamin A. JoyetLavergne has suggested that mitochondria contain a redox system in which vitamin A plays a part. A number of authors, Leblond, the present author and Giroud and his co-workers have demonstrated that mitochondria of some organs react with acetic acid-silver nitrate solution (which has been demonstrated as being a specific reagent for vitamin C) to give a positive reaction and this indicates that vitamin C may be present in them. One has to accept the intracellular localization of vitamin C with a certain amount of discretion in view of the very destructive effect of this reagent on the cell cytoplasm.

Electron-microscope studies in the present author's laboratory have demonstrated that this reagent has a most drastic effect on the ultrastructure of the cell and we cannot be sure that the localization of vitamin C in the mitochondria is, in fact, a real thing. It is of interest, however, that Chayen has found that the mitochondria of plant cells give this reaction very intensely and very specifically.

It may be that mitochondria in plant cells and some animal cells do, in fact, contain vitamin C, but we need further studies before this can be confirmed. Other vitamins appear to be present in mitochondria, members of the vitamin B complex have been found to be present, in some cases, in the form of coenzymes. The actual vitamins of the B complex recorded are vitamin B1 (thiamine), riboflavin, nicotinic acid (niacin), pantothenic acid and pyridoxine.

However, although these vitamins are present in mitochondria, they are not present in any greater concentration than in the other parts of the cell so that they are not exclusively contained in these organelles. These studies were made on cell homogenates and, of course, it is possible that the presence of the vitamins in the other fractions of the homogenate may be due to the fact that they have been leached out of the mitochondria by the saline solution which is used in the homogenization process in these particular experiments and further work would have to be done before we could be certain about this.

The present author, Joyet-Lavergne and Giroud have demonstrated that mitochondria contain glutithione or protein-bound SH and this is further evidence that mitochondria can play an important part in oxidation-reduction mechanisms in the cell. It is of interest that mitochondria have been found in large quantities in the phloem cells of plants which are concerned with transport and may be concerned with this process. This suggestion is made in view of Conway's views that redox systems can play an important part in ion pumping.

Another substance which was demonstrated histochemically in mitochondria by the present author, using the Schultz reaction, was cholesterol. This was particularly evident in the mitochondria of cells of the adrenal cortex. Similar but less intense reaction was shown by the mitochondria of the liver and it has been demonstrated that isolated liver mitochondria contain about 2% cholesterol and it is possible they contain more in the adrenal cortex.

Bensley has recorded the presence of a red pigment in the mitochondria and also in the submicroscopic particles of the liver cell. He believed that this pigment is derived from the oxidation of unsaturated fats and possibly the phospholipids of the mitochondria and this led him to suggest that in the liver cell the mitochondria in particular are, possibly, the seat of highly active oxidative processes which involve the metabolism of fats.

The possible relation of mitochondria to oxidation-reduction processes was indicated by the publications of Ludford. He demonstrated that, if methylene blue was added to tissue cultures, the mitochondria of the cells stained a brilliant blue colour but this could be inhibited by potassium cyanide. If the cells were exposed to a bright light, the blue colour was rapidly bleached.

It is of interest that the mitochondria, although they have

been demonstrated to contain a good deal of fat and lipid, do not give a positive reaction with Sudan III. They contain protein (quite a high proportion of it), but it is of interest that the earlier workers, using Millon's reagent, produced a negative reaction for protein.

Subsequently Bensley and Gersh using a Millon's reagent of a different formula were able to obtain positive results from the mitochondria of many tissues and they found that they were particularly well stained in frozen, dried sections and particularly in those of *Amblystoma* liver. The same authors demonstrated that, in undenatured, frozen, dried sections of *Amblystoma* liver, the mitochondria were destroyed if the sections were exposed to artificial gastric juice and artificial pancreatic juice.

Despite all this chemical information most of which was in existence by the beginning of the 1940's, there was no certainty as to the function of mitochondria and it was not until 1946 and 1947 that the late George Hogeboom and his colleagues at Bethesda completely revolutionized our ideas of their function by demonstrating that the major proportion of the succinic dehydrogenase and an appreciable proportion of the cytochrome oxidase activity of the liver cell were present in the mitochondria.

This at once suggested that these organelles were the major sites of aerobic respiration in the cell. It is of interest that as long ago as 1915, Dr. Kingsbury had suggested that mitochondria were concerned with the respiration of the cell. His reasons for this were largely due to his observations that anesthetics such as ether and chloroform, which depressed cellular respiration and the respiration of the animal in general, also broke up mitochondria in the cell.

The work of Hogeboom and his colleagues was carried out on homogenates of liver which had been produced by grinding up liver with saline. This gave poor preservation of the form of the mitochondria and subsequently 0.8 *M* sucrose was used–this preserved the nature and form of the mitochondria very well. In the beginning there was some doubt as to whether the material being assayed was in fact mitochondria or not, but when sucrose was used this doubt gradually disappeared. Eventually electron-microscope studies of the structure of the granules isolated by homogenization and differential centrifugation demonstrated beyond doubt that they contained the same structures as the mitochondria of normal cells.

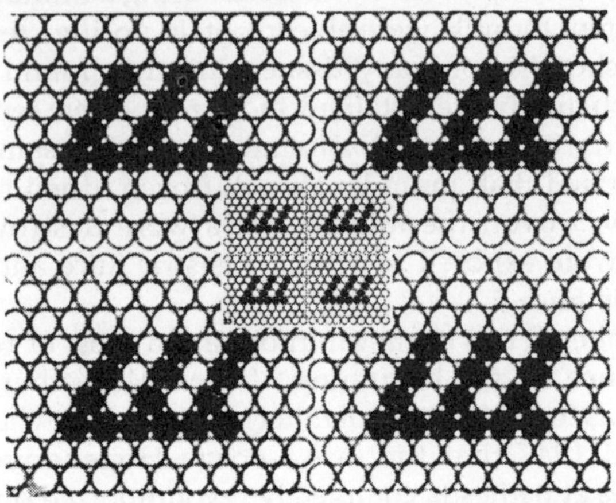

Fig. Localization of Respiratory Chain Enzymes (black spheres) in the Mitochondrial Membrane

Another possible source of error, however, began to haunt the biochemists and was frequently verbalized by cytologists and this was that the mitochondria did not really contain the oxidative enzymes, mentioned in the foregoing, but that they were being absorbed or adsorbed by them from the homogenate. This possibility was negated at least in part by adding more enzyme to the homogenate and demonstrating that it could be recovered almost 100% from the supernatant and so was not taken up by the mitochondria. Thus the localization of the enzymes in the original cell was probably in the mitochondria.

Subsequently, Kennedy and Lehninger demonstrated that in addition to containing cytochrome oxidase and succinic dehydrogenase, mitochondria also catalyze the condensation of pyruvic acid with oxaloacetic acid, the formation of succinic acid from ketoglutaric acid and the formation of malate from succinate. These three reactions represent three steps of fundamental importance in the Krebs tricarboxylic acid cycle of which we will have more details shortly. Mitochondria also contain coenzyme I of which nicotinamide is an important constituent and cytochrome reductase which is a flavoprotein. These two enzymes are links between the Krebs tricarboxylic acid cycle and the cytochrome system and it, therefore, appears that mitochondria probably contain the whole enzymic equipment necessary for aerobic respiration of the cell. In fact, it was originally demonstrated that a centrifugate of cells which included only nuclei and mitochondria were capable of carrying through the whole of the oxidation of glycogen to CO_2 and water; subsequently it was demonstrated that isolated mitochondria on their own could do this. Transaminase activity was subsequently demonstrated in mitochondria although in only the same concentration as in the rest of the cytoplasm.

Transaminase is concerned with protein synthesis and it is of interest that pyruvic, oxaloacetic and á-ketoglutaric acids are the corresponding á-keto acids of the amino acids, alanine, aspartic and glutamic and need only to be transaminated to produce them. Since the former compounds are themselves formed as intermediates in the course of the Krebs cycle, this indicates a mechanism whereby mitochondria might synthesize fresh protein material and so increase in size themselves or synthesize protein for other parts of the cell.

This possibility is further extended by the fact that RNA is known to be present in varying amounts in mitochondria, although the exact amount is uncertain and there is a possibility of contamination of mitochondria with RNA from the rest of the homogenate when making such estimates. However, there is no doubt that mitochondria do have the equipment for synthesising protein although to what extent they do this *in vivo* is not known.

All the fatty acid oxidase activity in the cell is also in the mitochondria and so is 80% of the octanoxidase activity. Since these discoveries, an enormous amount of work has been done on the mitochondrial enzymes and this work was summarized by Hogeboom (unfortunately now deceased), Kuff and

Schneider from the *National Institutes of Health* and has been published in Volume 6 of *The International Review of Cytology*, 1957. It may be of interest to record here other enzymes and compounds which they listed as being present in mitochondria.

It is of interest too, that a considerable amount of ribonuclease and deoxyribonuclease are present in mitochondria–this seems unusual since only a very small amount of DNA and a relatively small amount of RNA are present in the mitochondria compared with other parts of the cell. However, it is possible that the bodies which contain RNAase and DNAase are not true mitochondria but are lysosomes; we will discuss these shortly. In the meantime, it is of interest to note that it was eventually demonstrated by fragmenting mitochondria and separating the membranes from the contents that cytochrome oxidase and succinic dehydrogenase activity were located primarily in the membranes.

Subsequently it has been shown that the ability of the mitochondria to carry out the Krebs cycle and to combine this with oxidative phosphorylation is dependent on the presence of intact double membranes in the mitochondria. Now, it has been demonstrated and mentioned earlier that in the structure of mitochondria there is an outer membrane and an inner membrane to the mitochondrion, so that it is surrounded by a double membrane.

Also, the inner membrane is folded inward to give elongated cristae and these cristae, because they consist of a fold of the inner membrane, are themselves double membranes. There appears to be little difference biochemically between the double membrane structures of the cristae and the double membrane structure of the outer wall of the mitochondria.

An important factor in the process of oxidative phosphorylation is the transport of electrons. Dr. David E. Green has described the process of electron transport as carried out fundamentally with the aid of an electron transport particle–this is a submitochondria particle.

The latter can be isolated in a form so that it can no longer carry out the full citric acid cycle, but it can still carry out the process of oxidative phosphorylation which it performs by the oxidation of succinate and DPNH (reduced diphosphopyridine nucleotide, a codehydrogenaste) with freshly prepared by homogenization and differential centrifugation and are then exposed to alternate freezing and thawing, their form undergoes considerable change.

First of all, the external membrane ruptures and some of the cristae break off and eventually the mitochondrion is completely disintegrated; however, fragments of double membrane structure are left which, according to D. E. Green, may be derived either from the cristae or from the outer envelope of the original mitochondrion. These double membrane fragments represent collections of Dr. Green's electron transport particles.

If other methods are used for preparing these particles and the particles so produced do not show a double membrane structure, then oxidation of succinate and DPNH coupled with phosphorylation does not occur. Apparently the double membrane structure is essential for this process to take place. In the structures which do not have the double membrane structure, the capacity for electron transport is still present but oxidation phosphorylation cannot take place.

About 35% of the dry weight of the electron transport particle of Green is composed of lipid and the lipid is concentrated in packets of lipoprotein. These are interspersed between two molecules of cytochrome or between a flavoprotein and cytochrome. In other words, all the oxidation reduction members of the election transfer chain are connected to one another by lipoproteins.

These lipoproteins, according to Green, act not only as structural devices, holding various enzymes together, but they also conatin a compound which can shuttle electrons back and forth so that then can pass from cytochrome through to flavoprotein and reverse. This compound, which Green and his colleagues have described, is a new coenzyme called coenzyme Q.

It is a completely water insoluble benzoqtuinone derivative, which is connected in the lipoproteins of the electron transport particles and is related to vitamin K. Coenzyme Q can undergo oxidation and reduction in a similar fashion to cytochromes; for instance, the quinone part of the molecule can be reduced to become a hydroquinone and the hydroquinone can be reoxidized.

It is pretty certain that this coenzyme plays an important part in the electron transfer chain of the particle. The significance of the electron transport in these mitochondrial membranes may seem obscure but, as Green points out, one should think of this problem of electron transport in terms of its coupling–the fact that the process can be coupled to the esterification of inorganic phosphate. In this process a monophosphoric ester of the hydroquinone or, what Green described as a semiquinone form of coenzyme Q would be formed.

This ester would then react with adenosine diphosphate by a process of transphosphorylation and ATP would be forme and simultaneously the semiquinone or hydroquinone would be oxidized to the quinone form in coenzyme Q by the ferric form of cytochrome c. The production of ATP is, in fact, the reason for the existence of such a system.

This explanation of the way in which the oxidative processes of the mitochondria actually are used to produce ATP by oxidative phosphorylation is an extremely fundamental discovery and Green and his co-workers are to be complimented very much on the extraordinary patience and care which they have put in over a large number of years to work out this very complex problem in mitochondrial physiology.

In view of the studies by Green and his colleagues which have just been described, it is of interest to consider some of the studies of Martias. This author has for some time put forward the view that there may exist in mitochondria two routes for the transport of hydrogen ions or electrons, respectively, between the pyridine nucleotides (DPN and TPN) and cytochrome c. He thought that only one of these would be linked with enzymes which brought about phosphorylation. The route which is best known leads from DPN to cytochrome c reductase and then to cytochrome c. Martias believes that this particular route is not one which is concerned with the formation of highenergy phosphates.

On the other hand, he proposed a second alternative route in which vitamin K (phylloquinone) is an important factor. Until recently this concept of an electron transport system involving vitamin K in mitochondria has not proved to be particularly popular, Martias and Strouff have finally isolated a phylloquinone reductase that is a flavoprotein having a number of similarities to cytochrome c reductase.

There is, however, a, difference between them. Phylloquinone reductase and cytochrome c reductase can both react in the reduced state with phylloquinone, but with cytochrome c only cytochrome c reductase will have any effect. Vitamin K reductase has an extremely high activity even in relatively low concentrations.

If we consider that there are two routes in mitochondria which can be used for electron transport, then the problem to be solved is what decides whether these electrons and hydrogen atoms are directed into either of these particular pathways. Under normal circumstances they seem to be directed exclusively along the pathway which is coupled with oxidative phosphorylation and which leads via vitamin K reductase to vitamin K, cytochrome b and so through to cytochrome c.

This pathway via vitamin K seems to be sensitive to any alteration of the internal structure of the mitochondria, possibly because some of its constituents, particularly the vitamin K, are associated with the double membranes of the cristae. If their combination is disturbed, the electrons take the emergency route which is via cytochrome c reductase to cytochrome c. It is of interest in connection with this theory that dicoumarol is capable of decoupling oxidative phosphorylation from respiration and, since dicoumarol is known to be a vitamin K antagonist, its action is possibly explicable by the fact that it blocks the pathway between vitamin K and its reductase. It is also of interest that thyroxine is a hormone which causes uncoupling of respiration from oxidative phosphorylation. Since it is known that, if thyroxine is added to mitochondria, it causes a swelling and change in the internal structure, this would be perhaps further evidence that the internal structure of the mitochondria is the one which is particularly concerned with the flow of electrons along the route leading to oxidative phosphorylation.

It may well be that hormones which affect respiration do so by affecting the structural nature of the mitochondria, apart from their effects on the endoplasmic reticulum which have already been mentioned. The precise relation between Green's coenzyme Q and Martias' vitamin K phylloquinone reductase is not absolutely clear, but it is very likely they are identical or closely related systems.

LYSOSOMES

De Duve in his studies on ultracentrifugation of cytoplasmic particles has demonstrated that some particles appear to contain acid phosphatase, cathepsin and ß-glucuronidase and ribonuclease, DNAase and cholinesterase. It is of interest that these compounds appear to be present in separate particles–they require higher centrifugal forces for sedimentation than do the cytochrome-oxidase bearing mitochondria. De Duve has pointed out that it is of special interest that such hydrolytic enzymes are in special particulate bodies differing from others of the cytoplasm.

He is not clear as to how this should be interpreted but has pointed out that, if hydrolytic enzymes are free to act within the living cell, as, for instance, they can do in homogenates, they would interfere with the efficiency of the synthesis and may even affect the structural integrity of the cell; he then suggests that segregation of hydrolytic enzymes in this way is one method by means of which this activity is either kept in check or localized in specific parts of the cell.

De Duve once described these bodies as "suicide bags" and suggested that on the death of the cell these enzymes are released completely into the cytoplasm and play a part in the process of autolysis. The present author has suggested that these enzymes may be less effectively contained in the lysosomes in senescence and that this leakage may be partly responsible for the cellular process of aging. It is of interest that lysosomes have been demonstrated *in situ* with the electron microscope and stained to demonstrate their acid phosphatase activity. No typical mitochondrial structure can be seen in these bodies and they appear to be of a different nature from mitochondria.

Chapter 3

Chemical Cell Metabolism

THE METABOLISM OF CARBOHYDRATES

The role of mitochondria in oxidative phosphorylation has already been mentioned and their role in the metabolism of carbohydrates through the presence in their substance of enzymes concerned with the Krebs cycle and the cytochrome system has been indicated. We should now consider the problem of carbohydrate metabolism and the role that mitochondria and other parts of the cytoplasm play in it.

Carbohydrate metabolism is extremely important for cell synthesis and is the main source of energy for cell activities. Before attempting to localize the various activities of carbohydrate metabolism in the actual parts of the cell, we should consider briefly what the metabolism of carbohydrates involves. There are two types of metabolism, anaerobic and aerobic. The anaerobic route is demonstrated very well by muscle and most of the information on this type of metabolism of carbohydrates has been obtained by studies of this tissue.

The result of anaerobic metabolism is the production of lactic acid and the liberation of a good deal of CO_2. However, although we think in terms of anaerobic metabolism for muscle, we have to realise that muscle itself has a first class blood supply which appears to be increased by various physiological mechanisms when muscle is forced to do work and that muscle in the process of contraction uses a rather surprisingly large amount of oxygen.

Lactic acid has been shown to accumulate in muscle extracts and in isolated muscles kept under anaerobic conditions. If we consider the accumulation of lactic acid in an animal in vivo, we find that after moderate work the accumulation of lactic acid goes up slightly but remains at a pretty steady level. However, if strenuous work is done, then the amount of lactic acid goes up extremely steeply and slowly comes back to normal.

The reason for this is that under normal circumstances muscle can obtain oxygen fast enough to reoxidize the lactic acid as rapidly as it is formed and only a small amount of lactic acid accumulates. However, it is possible for muscle to do more work than it can supply oxygen for and it can do this by oxidizing carbohydrates anaerobically and so accumulating lactic acid. Eventually this lactic acid has to be converted with the aid of oxygen but this can take place over a longer period. Some of the lactic acid is converted into glycogen in the liver and the rest is oxidized.

The fact that it is possible to accumulate lactic acid and slowly oxidize this after the work is finished provides a mechanism by means of which an "oxygen debt" can be produced. Under aerobic circumstances it is not lactic acid which is formed in the metabolism of carbohydrates but pyruvate. However, this does not accumulate and it is oxidized almost as rapidly as it is formed–as we shall see in a minute there is a very complicated system for oxidizing this pyruvate.

The only time when pyruvic acid does accumulate in the tissues is in the absence of vitamin B_1 or thiamine, a fact which was demonstrated years ago in Oxford by R. A. Peters. Under anaerobic conditions pyruvic acid undergoes the process known as oxidative decarboxylation with the aid of the cocarboxylase (thiamine pyrophosphate). It yields acetate, carbon dioxide and lactate. It has been stressed that this reaction is fundamentally of an oxidative nature and leads to the term oxidative decarboxylation which is in important process both in carbohydrate and protein metabolism.

We have mentioned the production of pyruvate but have not yet considered the process by means of which this compound is produced–this process is known as glycolysis and it represents the first stage in the metabolism of carbohydrates. Glycolysis is, in effect, the reverse of photosynthesis. In photosynthesis the energy of sunlight is used to combine CO_2 and water into carbohydrates; in the process of glycolysis, the glucose which is formed from carbohydrates such as glycogen and other polysaccharides is converted into CO_2 and water with the liberation of energy.

We can express this as:

$$C_6H_{12}O_6 + 6O_2 = H_2O + 6CO_2 + \text{energy.}$$

This oxidation of glucose to give CO_2, water and energy is not a single step but involves a very large number of steps of considerable complexity.

During the various steps, small packets of energy are released at a rate at which the cell can use them, whereas if there was a sudden explosive release of energy by the oxidation of glucose the cell would probably not be able to use this relatively large amount of energy in a coordinated way and a good deal of it would probably be wasted. The first stages of glycolysis involve the phosphorylation of glucose and this is done with the aid of ATP as follows (the enzyme concerned in this process is placed above the arrow):

$$\text{glucose} + \text{ATP} \xrightarrow{\text{hexokinase}} \text{glucose-6-phosphate} + \text{ADP}$$

Hexokinase is not just one enzyme, there are in fact a number of hexokinases that catalyze phosphorylation of hexoses. The phosphorylation of glucose results in the transferring of a high-energy phosphate from the ATP to glucose and so to form a phosphate ester which is poor in energy–this type of reaction is called an exergonic reaction and is essentially irreversible.

It is of interest that one of the properties of glucose-6-phosphate which differs from glucose is the fact that the phosphate ester has difficulty in penetrating cell membranes whereas glucose itself, apparently, crosses without any difficulty and it has been suggested that this hexokinase reaction is one way in which glucose can be locked in a cell. It is also an essential prerequisite for the resynthesis of glycogen.

Many things can happen to glucose apart from

phosphorylation and ultimate conversion into CO_2 and water via the glycolytic and Krebs cycle system. Amongst these are its dehydrogenation by a glucose dehydrogenase to form gluconic acid. This is done by a diphosphopyridine nucleotide (DPN) linked (codehydrogenase) reaction.

In mammals, it is believed that this is not a usual pathway for glucose to follow. If glucose can be locked into position in a cell by being converted into a phosphate, it is obvious that there must be in existence a mechanism which can release it again since it needs, for instance, to be fed from the liver periodically into the blood stream to keep the blood glucose level at a relatively constant figure. This is carried out by a specific enzyme, glucose-6-phosphatase,

$$\text{glucose-6-phosphate} + H_2O \xrightarrow{\text{G-6-Pase}} \text{glucose} + PO_4.$$

This glucose-6-phosphatase is probably present in all tissues which release glucose from cells, but it does not appear to occur in skeletal muscle.

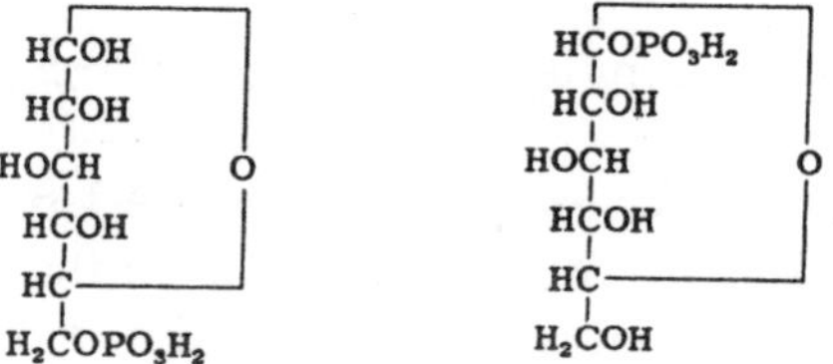

Fig. Glucose-6-phosphate α-Glucose-1-phosphate

Glucose-6-phosphate may be converted into glucose-1phosphate, in other words the phosphate is simply shifted around from the 6-position to the 1-position on the glucose molecule.

This change of position of the phosphate group is catalyzed by an enzyme known as phosphoglucomutase and the reaction changing the phosphate group from one part of the molecule to the other is an easily reversible reaction. The formation of glucose-1-phosphate from glucose-6-phosphate is a stage in the synthesis of glycogen, for instance, if glucose is accumulating in a cell it is converted to glucose-6-phosphate and then to glucose-1-phosphate.

The molecule of glucose-1-phosphate then polymerizes into glycogen. The reverse process can occur–glucose-6-

phosphate can be obtained from glycogen by first producing glucose-1phosphate and then by producing glucose-6-phosphate from the glucose-1-phosphate. We have, however, diverted a little from the direct line of our story of the glycolytic cycle.

After the formation of glucose-6-phosphate from glucose, the next stage in the glycolytic cycle is the conversion of glucose-6phosphate into fructose-6-phosphate. This reaction is readily reversible and is catalyzed by an enzyme called phosphohexose isomerase.

Fructose-6-phosphate may also be formed directly from fructose and it is known that an enzyme, fructokinase, which is present in brain and muscle can produce this effect. The next step is the further phosphorylation of fructose-6-phosphate; another phosphate group is added in the 1-position to give fructose 1-6-diphosphate. The enzyme responsible for this is phosphofructokinase and the reaction is carried out with the aid of ATP:

fructose-6-phosphate + ATP $\xrightarrow{\text{PFkinase}}$ fructose-1-6-diP + ADP

This reaction is also exergonic: Fructose-1-6-diphosphate is also known as hexose diphosphate and in the next stage this is broken down by what is described as the "aldolase" reaction into 3 phosphorylated compounds–triosephosphates. The first of these is ketose triosephosphate, the second is dihydroxyacetone phosphate, the third is phosphogly ceraldehyde. The triosephosphate can be converted into phosphogly ceric acid; dihydroxyacetone phosphate can be converted into phosphoglyceraldehyde or alternatively it can be reduced to glycerophosphate with the aid of DPNH and á-glycerophosphate dehydrogenase.

This reaction is potentially important for the synthesis of lipids since from the á-glycerophosphate, phosphatidic acid can be synthesized and phosphatidic acid can be the starting point for the synthesis of lecithin, cephalin and also of fats. Phosphoglyceric acid with the aid of the enzyme enolase becomes converted into phosphoenol-pyruvic acid.

Pyruvic acid penetrates cell membranes very well and can

thus leave the cell and in theory can be distributed in any cell in the body. Conversely all the steps which we have mentioned can be reversed and pyruvate can be converted back into glucose-6-phosphate. When pyruvic acid is reduced it gives lactic acid. Lactic acid itself can be converted back to pyruvic acid. Liver cells are capable of reversing the whole glycolytic series of reactions and can produce glucose and glycogen again from lactic acid. Muscle can reverse lactic acid to glucose-6-phosphate and glucose-1-phosphate and glycogen but cannot produce nonphosphorylated glucose.

In these first stages of carbohydrate metabolism, a certain amount of energy is liberated but this is not much more than about one-tenth of the total amount of energy which is produced by the production of CO_2 and water from glucose. The glycolytic part of the cycle is the least energy producing part. In addition to its conversion into lactic acid or its oxidation, pyruvic acid can be converted to alanine, an amino acid, by the process of transamination. Thus here one can see a link between carbohydrate and protein metabolism.

If there is ample oxygen, the pyruvic acid produced by this first stage of carbohydrate metabolism can be oxidized. This is a complex process in which in the first stage the pyruvic acid is converted by the process of oxidative decarboxylation into acetyl coenzyme A and CO_2. Thiamine pyrophosphate (cocarboxylase) is an essential enzyme for this process.

Acetyl coenzyme A is a 2-carbon substance and it condenses with oxaloacetic acid which is a 4-carbon dicarboxylic acid to yield a 6-carbon tricarboxylic acid, namely, citric acid. The enzyme catalyzing this is called the "condensing enzyme." Thus starts a series of changes which ultimately lead to the formation of CO_2 and water. The production of citric acid is followed by the loss and recapture of water and it becomes converted to cis-aconitic acid (with the aid of aconitase, glutathione and ferrous iron) which on further hydration is converted into isocitric acid.

This compound then loses hydrogen and thereby becomes oxidized to oxalosuccinic acid (this reaction is catalyzed by isocitric dehydrogenase) and decarboxylation turns it into á-ketoglutaric acid (the enzyme responsible is oxalosuccinic decarboxylase and oxidized manganese). á-Ketoglutaric acid is decarboxylated and then oxidized by the loss of two hydrogen atoms to succinic acid (the enzyme concerned is á-ketoglutaric dehydrogenase).

Succinic acid is also oxidized by the loss of H_2 (with the help of succinic dehydrogenase) to fumaric acid. The latter by the addition of the elements of water becomes converted by fumarase into malic acid and the malic acid by dehydrogenation (enzyme malic dehydrogenase) is converted in oxaloacetic acid and there we are back at the beginning of the cycle again. The oxaloacetic acid is ready to combine with another molecule of acetyl coenzyme A to produce citric acid once more. During this process 3 molecules of CO_2 are given off and 5 molecules of H_2.

It can be seen that quite a complex series of reactions take place in what has been called the "tricarboxylic" or "citric acid cycle," or the "Krebs cycle." All three terms are applicable. The final combination of the hydrogen atoms liberated by the Krebs cycle with oxygen is brought about by the cytochrome system. It is of interest that cytochrome is a protein which contains a form of heme, an iron containing pigment which is also present in hemoglobin.

It is probably more widely distributed than any other type of heme protein since it occurs in the cells of all organisms that use oxygen, irrespective of whether they are animals or plants or whether in the case of animals they are vertebrates or invertebrates, or protozoa.

In the cytochrome system a variety of compounds and enzymes are involved. Cytochrome oxidase is a widely distributed enzyme since its occurrence is comparable in distribution to cytochrome. It has not yet been obtained pure and it has been found to be bound to the insoluble material when cells are homogenized and spun down.

This is not surprising when we appreciate that cytochrome oxidase is associated with the mitochondria. There is also another compound involved called flavoadenine dinucleotide (FAD) for short) which is associated with various proteins to form a variety of different enzymes. A flavine containing enzyme was first isolated and called a yellow ferment as long ago as 1932 by Warburg and Christian.

Flavoproteins are usually associated with metals of various sorts and molybdenum, copper and iron are three which have been found to be necessary for their adequate functioning. It is not quite clear where the metal atoms are situated on the molecule but they appear to be essential for the action of the enzymes with which they are associated. There are two types of riboflavin containing enzymes, one type is an electron acceptor from reduced DPN or TPN and it can transfer these either to oxygen or to the cytochromes, whereas the other type of flavoprotein accepts electrons directly from metabolites.

Among the important flavoproteins is cytochronie reductase which comes in two types–one for reduced DPN (DPNH) and the other for reduced TPN (TPNH). The oxidation of these two compounds (DPNH and TPNH) in the cell can thus be carried out by the cytochrome system. Now we are in a position to describe the next stages which take place in the oxidation of glucose.

The 5 pairs of hydrogen atoms which are passed down to the cytochrome system from the Krebs cycle are combined with the cytochrome and result in its reduction. The enzyme concerned is cytochrome reductase, the flavoprotein already mentioned. Cytochrome is then oxidized with the aid of cytochrome oxidase, which removes the hydrogen atoms. They become combined with atmospheric oxygen to produce water and thus the long journey of oxidation of carbohydrates is done, 5 molecules of water and 3 molecules of CO_2 being produced from each molecule of pyruvate.

An analysis of oxygen uptakes of various tissues in the body gives an indication of the degree to which their cells are metabolizing. As a matter of interest, it might be noted that of the tissues examined, retina and kidney had the highest metabolism and liver was next, then the rate decreased progressively from adrenal, lung, bone marrow, diaphragm, heart, lymph nodes, skeletal muscle, skin to eye lens which had the lowest level of oxygen consumption. The important point to remember is that the complex of reactions described above is localized to a great extent in the mitochondria.

One of the important functions of this oxidative cycle just described is its relationship to phosphorylation. We have described earlier the work of Green who showed that oxidative phosphorylation could take place without the whole Krebs cycle occurring, but in intact mitochondria the whole cycle normally goes through and phosphorylation is an important by-product of these reactions. This process of phosphorylation results in the production of high-energy phosphate esters, such as ATP. Since ATP is one of the principal energy containing compounds of the body, it is very important in the whole energy cycle of the cell.

It is, for example, the main source of energy in muscular contraction and for many of the synthetic processes of other cells. This briefly is the story of carbohydrate metabolism in the cell. What we want to try to do now is to demonstrate where this complex of activity is situated in the living cell. It appears that most of the processes of respiration and glycolysis actually take place in the cytoplasm and mitochondria.

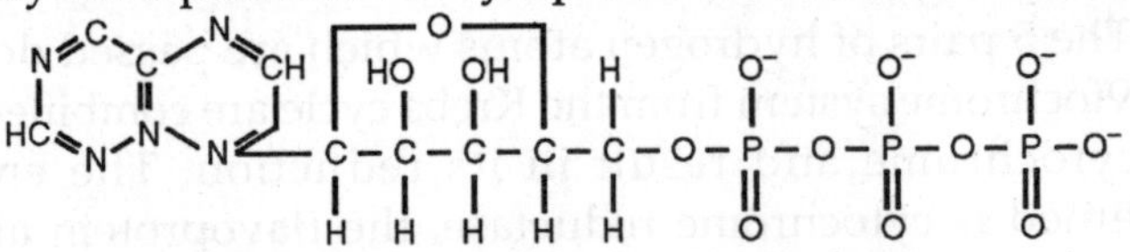

Fig. Adenosine Triphosphate (ATP)

Some years ago, the present author and R. J. Allen demonstrated that zymohexase which is really a complex of two enzymes and is concerned with the splitting of hexose diphosphate (the aldolase reaction) was localized in the cytoplasm of the cell and in the case of muscle fibers in between the fibrils in the sarcoplasmic material rather than in any of the formed elements.

It is noted too that the results of differential centrifugation of cell homogenates have demonstrated that most of the enzymes responsible for the glycolytic cycle have been found to be present either in the supernatant or in "microsomes" and those concerned with the cytochrome system and Krebs cycle ire localized specifically in the mitochondria. Now, since the microsomes mostly represent fragments of the endoplasmic reticulum, we can assume that most of the enzymes found in them can also be found in the membranes of the endoplasmic reticulum.

In addition to the aldolase complex which has been demonstrated to be present in the supernatant, it has been found that phosphorylase and phosphoglucomutase and glucose 6-phosphate dehydrogenase have also been found in the supernatant and glycolysis has in fact been found to take place in this fluid–a fact which confirms that all the glycolytic enzymes necessary for this process must be present there.

However, glucose-6-phosphatase has been found not to be localized at all in the supernatant but exclusively in the microsomes (fragmented endoplasmic reticulum). There is some evidence too that hexokinase is present in the microsomes. On the other hand, DPNH and DPN and TPN-linked cytochrome c reductases which are found in high concentration in mitochondria have also been found to be present in the microsomes. It might be possible to suggest from these facts a tentative scheme which would help us to understand to some extent the relationship between the reactions occurring in the cytoplasm itself and those in the mitochondria.

Let us assume that glucose passes across the cell membrane, perhaps it is taken in by pinocytosis as has been suggested by a number of authors or, possibly, if and when the membranes of the reticulum are continuous with the outside of the cell, it may simply pass up the cisternae between the membranes of the reticulum and enter into the cell through the membranes of this reticulum.

Let us assume that the glucose has passed into the cell by the process of pinocytosis. Then, with hexokinase present in the cytoplasm, it can be converted into glucose-6-phosphate and subsequently run through the cycle to pyruvate.

At this point (another system which we will discuss in a minute) associated with the mitochondria comes into play. If the glucose has to pass across the cell membrane or across the membranes of the endoplasmic reticulum, it should also be able to do this without too much difficulty. If any glucose-6-phospate should occur in the cisternae it will probably be able to pass freely into the cytoplasm because of the presence of glucose-6-phosphatase in the membrane. It may be that there is hexokinase in the fluid within the cisternae and that, perhaps, all the glucose there (if any) is first phosphorylated and then released in a timed fashion into the cytoplasm through the action of glucose-6-phosphatase in the membranes.

Perhaps this is a way of controlling the feeding of the carbohydrate fuel into the furnace. It will be remembered that glucose-6-phosphate passes cell membranes with difficulty unless there is the appropriate hydrolytic enzyme on the membrane. The glucose, once into the cytoplasm by whatever route, can be synthesized into glycogen or it can be brought to pyruvate by the enzymes of the glycolytic cycle which are known to occur in the cytoplasm. Once pyruvate is prepared, it is changed into acetyl coenzyme A and is ready for the next stage.

For this it has to come into contact with the mitochondrial membrane. Whether it has to pass through the mitochondrial membrane and come in contact with the cristae we do not know, but the existing evidence suggests that the envelope and cristae of mitochondria have fundamentally the same enzymatic structure. Furthermore, there are pores and even large openings in some mitochondrial envelopes and there is thus little difficulty in metabolites passing in or out.

Now, it has been mentioned that a great proportion of the enzymes concerned with the Krebs tricarboxylic acid cycle are localized in the mitochondrial membranes and since acetyl coenzyme A appears to be the most commonly used fuel of this system, it is at this point that the mitochondria largely take over the final oxidative stages. Thus, as far as we can tell at the moment either in the cytoplasm of the cell itself or in the endoplasmic reticulum, glycolysis occurs with the production of either acetyl coenzyme A or pyruvic acid. We do not know at what stage the latter are fed to the mitochondria.

They must go largely to the mitochondria because these organelles contain the greatest proportion of the Krebs cycle enzymes–it is these latter which complete the oxidative breakdown of carbohydrate that was set into motion by hexokinase. The rate of penetration of these fuels into, or their absorption onto the mitochondria will be conditioned both by the nature of the mitochondria and the total area of mitochondria available. If under normal conditions we have a large number of mitochondria occupying the cell, we can assume, other things being equal, that oxidation should be proceeding at a fast rate.

However, where one gets a large increase of number of mitochondria as, for example, in starvation this may be compensatory hypertrophy and not indicative of increase in oxidative activity. We do not know whether the nature of the mitochondrial surface varies from time to time but we know that the surface area varies.

Mitochondria, for instance, fragment from the filamentous and rodlike condition to the granular state under a variety of circumstances. These include mechanical damage to the cells, rough handling, influence of bacteria and other toxins, as a result of anesthesia and anoxia and so on. It can be shown mathematically that there is a greater surface area available the more fragmented the mitochondria become. Because of this either the pyruvate and/or acetyl coenzyme

A can feed more rapidly into or become attached in larger amounts to the mitochondria when they are in a fragmented condition simply because there is a greater surface area. Thus the mitochondria in virtue of their ability to fragment and reform into the filamentous condition can act as a throttle controlling the rate of aerobic metabolism in the cell.

They may also have an additional means of doing this, a sort of fine control. Since it is fairly certain that enzyme molecules are aligned along the cristae, these structures represent another surface where reactions can take place and reduction in the size and number of cristae would also have a throttling effect on the metabolism or synthesis due to mitochondria.

This can be seen in operation in the case of a mitochondrion which has accumulated a good deal of fat or other product of chemical reactions. In such a case, the cristae are reduced or absent altogether as if metabolic processes are brought to a virtual standstill because of the accumulation of reaction products. This brings us to another process of control, a chemical method known as "feedback" which will be discussed shortly.

Siekeiwitz believes that both glycogen synthesis and breakdown take place at the surface between the cytoplasm and the ergastoplasmic membranes in association with the enzymes present at those surfaces. He points out, however, that only glucose-6-phosphatase and DPNH and TPNH cytochrome c reductases (these latter are believed to act as coenzymes for glycolysis in the early stages of glucose oxidation) have been found to be associated with the microsome fraction, but he believes that the site of effective action of the other enzymes might be at the interface between the membranes and the matrix of the cytoplasm. He suggests that hexokinase might be activated at the membrane surface of the endoplasmic reticulum. Other cofactors such as glucose-1-6-diphosphate and adenosine monophosphate possibly also bind their appropriate enzymes to the E.R. membranes.

Siekewitz has also discussed certain biochemical aspects of the control of glucose metabolism. First of all he points out that, if the concentration of glucose in the lumen of the endoplasmic reticulum is in equilibrium with the glucose concentration in the blood (this assumes at least temporary continuity between the lumen or cisternae of the E.R. and the exterior of the cell), there exists then a mechanism whereby the glucose level in the blood would definitely affect intracellular glucose equilibrium.

Thus a reduced production of glucose from the diet would lead to reduction of glucose in the blood and this would cause a reduction of the amount of glucose in the fluid within the endoplasmic reticulum. The latter result would lead to increased phosphatase activity which would cause an increase in the breakdown of glycogen (e.g., in the liver cells) and a production of glucose which would pass out from the endoplasmic reticulum into the blood stream.

Siekewitz points out that hexokinase, phosphoglu comutase and phosphorylase might have their activity enhanced if they were attached to the endoplasmic reticulum membranes. In the case of phosphorylase, active phosphorylase B has to undergo a conversion to active phosphorylase A before it can have any catalytic effect. An enzyme phosphorylase B-kinase carries out this conversion by phosphorylating the enzyme in the presence of ATP. Siekewitz suggests that this kinase may be part of the membrane of the endoplasmic reticulum and that the activating process for phosphatase in the cell might consist of moving it out of the cytoplasm onto the site of the E.R. membranes.

Siekewitz points out that it is possible that hormones control this type of movement of enzymes within the cell and its internal membranes. At this point one might remember the work quoted earlier by Brandes and the present author in which it was demonstrated that shifts of acid phosphatase activity took place from the Golgi apparatus to the nucleus in ventral-lobe prostate cells following castration and that acid phosphatase activity was restored to the Golgi apparatus following implantation of the male sex hormone.

Also of interest are the further studies of Brandes in which he has demonstrated that in castrated animals there is a rearrangement of the membranes of the endoplasmic reticulum. These studies demonstrate a definite morphologico biochemical effect on the part of the male sex hormone. Siekewitz suggests that the hormones concerned with carbohydrate metabolism do not act directly on the enzyme as such but bring it and the substrate and various cofactors together at a suitable surface and then complex them together there. Other enzymes may be localized on the endoplasmic reticulum e.g. 5 nucleotidase.

The nucleus, as will be described later, appears to be enclosed in a fold of the double membrane of the endoplasmic reticulum and not to have a membrane of its own. Since the E.R. spaces might be connected directly to the exterior of the cell, it is possible that glucose coming from outside the cell could be converted into glucose-6-phosphate in the E.R. lumen and thus be prevented from passing through the E.R. membrane.

Thus it could pass along all the ramifications of the E.R. canals and come in direct contact with the nuclear fold of the reticulum. If this membrane contains glucose-6-phosphatase, it could penetrate through into the nucleus without having to pass through the cytoplasm at all. Histochemical preparations show that the glucose-6-phosphatase reaction in a particular histological section is not always positive for all nuclei in the section and it is of interest that Siekewitz has pointed out that enzymes may be present or activated at the E.R. membrane only when the glucose concentration reaches a critical level.

It is of interest that histochemical studies of the cells of many organs demonstrate the fact that many dephospho rylating enzymes as well as glucose-6-phosphatase show an association with nuclear membranes–perhaps here lies the mechanism whereby even low levels of glucose-6-phosphate could penetrate readily through into the interior of the nucleus. The endoplasmic reticulum could provide a pathway straight to the nucleus which would prevent glucose from getting in contact with or passing through the cytoplasm where it could be attacked by glycolytic enzymes. This may be the mechanism by which a supply of glucose to the nucleus is ensured. Glucose could also be supplied to the nucleus from the cytoplasm by the process of glycolysis.

In this case, glucose being formed from glycogen as glucose6-phosphate would be dephosphorylated and pass through the E.R. membrane and into the lumen where possibly it would be rephosphorylated to prevent its passing back again and would thus move along the lumen to the nucleus. Hence the nucleus could get its glucose directly from the cytoplasm via the endoplasmic reticulum or directly from the outside (the latter, however, only if a direct connection really exists).

In many cells the mitochondria can also be seen to be completely surrounded by endoplasmic reticulum. It is possible that the mitochondria themselves obtain their glycolytic fuel directly as a result of glucose passing through the E.R. membranes undergoing glycolysis there and the glycolytic products feeding directly to the mitochondria.

The control of the rate of different types of metabolism, particularly respiration, in the cell can depend on two main factors. First, a structural factor which brings into apposition the appropriate reactants and which varies with the extent of the surfaces available for these processes to take place and, second, it may also depend on some chemical feedback where an excessive production of one kind of compound inhibits its continued production or slows down further synthesis. Alternatively, the metabolism in any one particular direction may be affected by the absence of a limiting amount of some specific substance or compound in the reaction chain.

Sir Hans Krebs in a recent article entitled *"Rate Limiting Factors in Cell Respiration"* discussed the control of energy utilization and pointed out that, in unicellular organisms, energy can be obtained directly from oxidation if air is present but if air is not present then anaerobic fermentation takes its place and energy is obtained by this source. In this instance it is the supply of air or oxygen which regulates which of these mechanisms comes into use and oxygen is, in fact, the "rate limiting factor."

In higher animals the ability to undergo fermentation or an aerobic oxidation is still present and can be particularly well demonstrated in muscle. Krebs pointed out that the chemical systems in the cell which are concerned with the function of regulation are all fairly simple reactions but that there is an elaborate interlocking of these reactions. By this he means that the individual reactants may take part not only in more than one reaction but in very many different processes.

One of the difficulties in sorting out such a complex of activity is that not all the component reactions of this elaborate interlocking series are known and we are dealing with a heterogeneous system in which there are many varied membranes and different spatial arrangements of the various reactants. He also points out that regulation is probably a matter of reaction velocities, some of which will be accelerated and some slowed down and the question that has to be decided is the degree to which any of these are ratelimiting. Krebs illustrates this point by considering the amount of oxygen used by 4-ml. sheep heart homogenate, which contained about 10% of tissue.

Thus one can demonstrate that the addition of glycogen adds nothing to the oxygen uptake so the amount of glycogen present is not a limiting factor in this system. The same applies if glucose is added instead of glycogen. On the other hand, when acetate, pyruvate, or other intermediates of the tricarboxylic acid cycle are added there is an appreciable increase in the rate of oxygen uptake.

The fact that the oxygen uptake in this system can be increased if suitable substrates are added to the mixture demonstrates that the electron transport system from DPNH to oxygen is not being used to its full capacity, thus it cannot be the factor which is limiting the uptake of oxygen. Certain special substrates which are known to reduce either DPN or flavoprotein seem to be able to increase oxygen consumption.

Thus the limiting factor appears to be that, the mechanism for the transport of hydrogen from DPN or flavoprotein is not being used to its full capacity if these special substrates are not present. The limiting step therefore is really the first stage in the electron transport system.

If it is found that a particular substrate increases the rate of respiration this is due to the fact that the substrate reacts more readily or easily with DPN or flavoprotein than any endogenous substrate already present. This is the reason why pyruvate or á-ketoglutarate or succinate are responsible for the stimulation of respiratory rate in the experiment quoted. However, even if we accept this we are still faced with the identification of the factor that decides the rate of reaction between substrate and DPN or flavoprotein.

Studies with dinitrophenol which decouples oxidative phosphorylation from respiration is of interest and helps to throw light on this problem. Perhaps we should first say a word or two about this action of dinitrophenol. The uncoupling of oxidative phosphorylation from respiration has been compared to putting a car into neutral gear and still leaving the engine running. If the respiratory activities are regarded as the engine and the phosphorylation as the process of making the car go, then dinitrophenol uncouples the engine from the transmission of the car, the engine continues to turn but the car does not move; in the cell the respiration goes on quite happily but no ATP is formed. Normally ATP is formed from ADP and inorganic phosphate and thus the factor which limits the rate of oxygen consumption and oxidation of pyruvate in the normal system such as we have described is not really the amount of enzymes present but actually the level of either ADP or inorganic phosphate.

In the experiments carried out by Krebs, inorganic phosphate was present in a fairly good concentration and further quantities added to the system did not stimulate respiration. Therefore it is almost certain that the limiting factor must be the amount of ADP which is available.

These experiments demonstrate the type of chemical control which a single compound can exert on a whole chain of reactions. One should also remember that another mechanism, a structural one that controls the rate of respiration, is the rate at which glucose can enter the cell and this can be hormonally controlled, although the method of action of the hormone is not exactly known.

It is of interest that one of the factors which probably affects the rate at which glucose can enter the cell (if, in fact, the endoplasmic reticulum is continuous with the outside of the cell) is the degree of complexity of the endoplasmic reticulum.

If this structure develops many ramifications, as presumably it seems able to do in certain cells such as spermatocytes, as demonstrated by Fawcett, then the surface area available for glucose to enter into the cytoplasm of the cell and so be metabolized is enormously increased or, conversely, it may be decreased by a reduction in complexity of the reticulum. To return to the subject of respiration and ATP formation, the reaction for this process can be given as follows:

$C_6H_{12}O_6$ (glucose) + 6 O_2 + 38 ADP + 38 H_3PO_4 '! 6 CO_2 + 44 H_2O + 38 ATP

This is the general reaction and is a summary of all the complex intermediatary reactions which in the end simply produce carbon dioxide, water and ATP. Since, as Slater and Houlsman have pointed out, cells contain only relatively small amounts of ADP, as soon as it is all converted into ATP the process of respiration will stop–ADP is thus the limiting factor. However, when the cell is stimulated to do work there is a breakdown of ATP according to the formula given by Slater and Houlsman,

38 ATP + 38 $H_2O \rightarrow$ 38 ADP + 38 $H_3PO_4 \rightarrow$ work

and, since ADP is now being re-formed, respiration can go on so long as there is some left to be resynthesized into ATP. The addition of further ATP will, of course, keep respiration going.

It is of interest that, if mitochondria that have been separated from the cell by differential centrifugation are permitted to stand for some hours at room temperature, the phenomenon of uncoupling (which can also be brought about by dinitrophenol) of the oxidative phosphorolytic system from respiration takes place.

This type of mitochondrial preparation is described as "aging" mitochondria and it is tempting to speculate whether in senescing tissues there may not be a progressive uncoupling of respiration from oxidative phosphorylation or a progressive hydrolysis of ADP so that less and less of this becomes available for synthesis of ATP.

It has been possible to isolate from mitochondria which have been aged in this way, a heme compound which will actually produce this uncoupling reaction. It has been described and given the name "mitochrome" and is fundamentally a pigment.

However, there appears to be a lipid component in this "mitochrome" heme-protein preparation which is the actual factor responsible for uncoupling and the heme protein of the mitochrome is not, in fact, the uncoupling factor at all. Mitochrome is very similar in structure and form to cytochrome and is probably derived from it. Certain unsaturated fatty acids such as oleic acid are also found to be active as uncoupling agents and the lipid isolated from the mitochrome particle also appears to contain an unsaturated fatty acid. The fact that an uncoupling agent can be produced *in vitro* this way is of considerable importance since it seems possible that the formation of such a substance in mitochondria might take place in vivo and may itself function as a controlling agent for respiration and oxidation.

FAT METABOLISM

The general view of metabolism of fatty acids at present is that they are broken down beginning at the end of the chain where the carboxyl group is and then the chain is progressively degraded as two carbon pieces are removed by a process of oxidation. This type of oxidation of fatty acids is known as ß-oxidation and is so called because the fatty acid is attacked oxidatively at the ß-carbon atom in the first instance. Among the products of this oxidation is the formation of acetyl coenzyme A and acyl coenzyme A. The latter can be subjected to further oxidation with the production of more acetyl coenzyme A (CoA).

Although very little has been said about the mechanism of degradation of all these fatty acids, it is of interest that the enzymes which catalyze these reactions are all located in or on the mitochondria. The acetyl CoA can enter the Krebs cycle by condensing with oxaloacetic acid and the final oxidation thus follows the same path as the carbohydrates. The acetyl CoA formed from pyruvic acid (i.e., from carbohydrate breakdown) can be used to synthesize fats and likewise so can acetyl CoA produced as a result of protein metabolism. We see, therefore the reason why the Krebs tricarboxylic acid cycle has been spoken of as the meeting place of protein, fat and carbohydrate metabolism.

Where precisely in the mitochondria the enzymes responsible for ß-oxidation of fatty acids are centered is not known for certain. It is very likely that they are more associated with the outer membrane of the mitochondrion than with the cristae so that the problem of penetration of the fatty acid through the membrane of the mitochondrion does not become so important. On the other hand, there is some evidence that pores occur in the mitochondrial membrane and if this is the case it is possible for the long-chain fatty acids to pass into the interior of the mitochondria and to become subject to ß-oxidation at this site by enzymes located in the cristae.

Hoberman has shown that in the mitochondria, deuterium labeled DPNH, which has been reduced during the oxidation of fat in the organelles, is not available for reactions that take place in the outside cytoplasm. This suggests that the enzymes concerned with the oxidation are localized within the cristae. It is of interest that recent electron micrographs of the adrenal cortex have shown large areas of the surface where the mitochondrial membrane is incomplete and the interior of the organelle is thus open to the penetration of the largest molecules and even particles of fat. Novikoff's scheme for the differential distribution of biochemical activities in liver cells.

Fat is metabolized largely in the way already described, but the synthesis of fat is also an important part of the activity of the cell. Fat cells have an important mechanical function to perform and fat itself is a valuable reserve store of energy. Fats, fatty acids and phosphorylated fats (phospholipids such as lecithin) are also important structural units of cell, mitochondrial and other membranes and their synthesis becomes an important cellular activity.

Fatty acids are made up of long chains of carbon atoms, usually an even number, with a COOH group stuck at one end. Fat is formed from a fatty acid molecule by a combination between the latter and a molecule of alcohol such as glycerol. The link takes place through the COOH group which reacts with the OH of the alcohol to eliminate a molecule of water. Fatty acids may be short or long or intermediate chained, a typical short-chain fatty acid is acetic acid which has only 2 carbon atoms and a typical long-chain fatty acid is palmitic acid which has 16 carbon atoms.

In the synthesis of these long-chain fatty acids, the starting point appears to be acetic acid. This combines with CoA to form acetyl CoA. This condenses with CO_2 to form malonyl CoA (the coenzyme A ester of malonic acid). This latter compound then condenses with another molecule of acetyl CoA to form an intermediate substance which becomes reduced to form a 4-carbon fatty acid which condenses with a molecule of malonyl CoA to give a 6-carbon fatty acid and so the carbon chain is built up. Then, as described, combination of the appropriate fatty acid with an alcohol such as glycerol gives an ester known as a fat.

Phospholipins such as lecithin, which play a very important part in the structure of the cell membranes, also have to be synthesized by the cell. These compounds are not only esters of a fatty acid but are simultaneously esters of phosphoric acid. In this synthetic activity ATP plays an important part. It starts the ball rolling by phosphorylating one of the OH groups of glycerol. To the other two OH groups, CoA esters of palmitic acid are attached. At this point the phosphoric acid group that was put on in the first step drops off. The molecule then reacts with a compound known as cytidine diphosophocholine.

This compound, which is really a coenzyme, drops the diphosphocholine part of its molecule nicely into the spot on the OH group which had been vacated by the first phosphoric group and so the lecithin is formed. Other coenzymes are concerned with the production of the other steps in the synthesis but these will have to be studied in more specialized works of biochemistry.

Dr. D. E. Green has pointed out that one of the surprising things about the synthesis of fatty acids is that the synthesis stops at 16-carbons. It is rare to get 12- and 14-carbon chain fatty acids and 18- or more carbon chains scarcely ever form. What tells the cell to break off the synthesis at that point is certainly an intriguing problem.

Although one would expect that mitochondria would be the principal fatty acid synthesizers of the cell, the belief at the moment is that they are not the site of synthesis and that it takes place at the surface of the endoplasmic reticulum. However, in the production of fats from the fatty acids and especially in the case of lecithin synthesis where ATP is required, the mitochondria make a contribution to the synthesis because of their ability to produce the latter material. Mitochondria do play a direct role in fat metabolism but, this role as mentioned earlier in this chapter, appears to be in fat degradation rather than in fat synthesis.

PLANT CELLS

So far we have considered only animal cells and it is of interest that Hackett in the *International Review of Cytology*, has pointed out that in plant cells glycolysis involves the plastids, the soluble fraction of cells and possibly the nucleus as well as the mitochondrial enzymes. He points out that many enzymes which are involved in the Krebs cycle are not exclusive to the mitochondria and that hydrogen transfer is not confined to these organelles.

Mitochondria he says react with the nucleus in the process of phosphorylation, they react with the chloroplasts in photosynthesis and they react with the microsomes in protein synthesis in the endoplasmic reticulum. He believes that a close relation between the cell membrane and the mitochondria may play an active part in the movement of substances into the cell or possibly in the growth of the cell wall of plants. Furthermore, he reminds us that the real unit is the cell itself, its various parts work as an integrated unit and one should only break them down for the purpose of trying to analyse the various processes in which they participate.

MITOCHONDRIA: POWER PLANT OF CELL

For thousands of years, men puzzled over the question of where body heat comes from. This heat was recognized as nearly synonymous with life itself–but what produced it? The source of the energy of man or beast was equally mysterious. In recent decades it has become clear that both heat and energy can be traced to some sausage-shaped organelles in our cells, the mitochondria.

Of course the ultimate source of energy for all plants and animals is sunlight. But the sun's energy can be harnessed by plants, through photosynthesis and stored in molecules of carbohydrates. When animals eat these carbohydrates and break them down to carbon dioxide and water, with the help of oxygen and an arsenal of enzymes, large amounts of energy become available. Animals immediately convert this energy into molecules of high-energy ATP (adenosine triphosphate)– the universal currency of energy in living things. Excluding only the very first stages in carbohydrate breakdown, which are called glycolysis, the entire, complicated process of energy transfer to ATP takes place within the mitochondria.

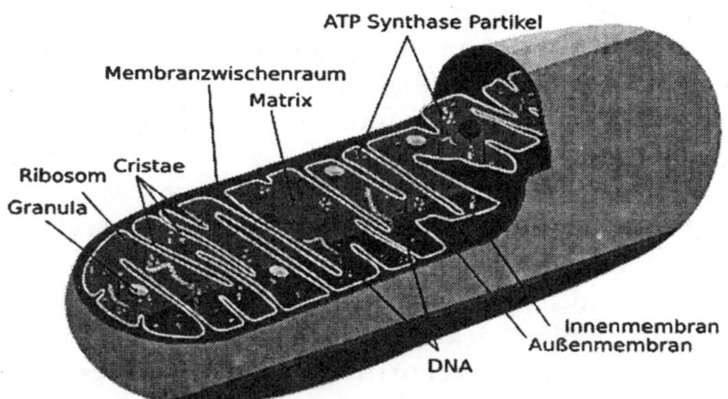

Fig. A Mitochondrion

Mitochondria are the largest organelles in the cell, after the nucleus, yet some cells have more than a thousand of them. Thin and long, they vary in diameter from 0.5 to 1 micrometer and in length up to 7 micrometers. Their thread-like outlines can be seen with a good light microscope. Although they were first observed in 1850, it took a century to visualize their internal structure and understand their function.

Actually the story goes back even further, to the 18th century, when Lavoisier showed that animals require oxygen for respiration. This led chemists to study cellular respiration, the process by which living cells use oxygen to release the chemical energy stored in foodstuffs. Early in the 20th century biochemists discovered that these reactions fall into two major groups:

- The carbon pathway, where a series of chemical

reactions, each requiring a specific enzyme, breaks down the carbohydrates into carbon dioxide and hydrogen;

- The hydrogen pathway, which transfers the hydrogen to oxyen in stages, forming water and releasing energy.

The whole process is organized much like an assembly line. In the hydrogen pathway, the hydrogen's electrons pass through an "electron transport chain" made up of enzymes that act as carriers and as the electrons move from enzyme to enzyme, they give up part of their energy, which is stored in molecules of ATP. The ironcontaining enzymes, cytochrome a, b and c, carry out the final stages of the process. All in all, three molecules of ATP are formed for every atom of oxygen that is used up in respiration.

For a long time, biochemists studied these reactions in cell extracts without worrying about what parts of the cells might be involved. One scientist, B.F. Kingsbury of Cornell University, did suggest that mitochondria might be the site of cellular respiration back in 1912, but this was ignored. In the 1940's, using an ultracentrifuge, Albert Claude and his associates isolated various cell particles including mitochondria. Shortly afterwards, in 1948, Albert Lehninger of Johns Hopkins and Eugene Kennedy of Harvard showed that the reactions leading to the synthesis of ATP occurred in the mitochondria.

In the early 1950's, Palade and the Swedish scientist Fritiof Sjostrand reported that mitochondria are bounded by a membrane and that they have a system of parallel, regularly spaced inner ridges which were named cristae. It is now clear that there are, in fact, two membranes around the mitochondria of plants, protozoa, molluscs, insects and man: an outer membrane, set off by a fluidfilled gap; and an inner membrane which is folded inward at various points to increase its surface, forming the cristae, so that the enzyme molecules of the electron transport chain, which are attached to this inner surface in specific sequences, can be packed more densely side by side in the mitochondria.

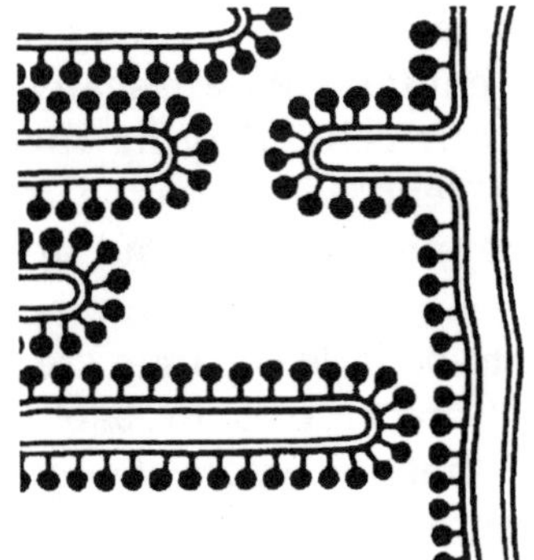

Fig. The Intricate Internal Structure of a Mitochondrion, with Outer Membrane and Particles on the Folded Inner Membrane, Part of which forms the Cristae

This general design for mitochondria seems to have existed unchanged, through the evolution of plants and animals, for more than 1.2 billion years. Though mitochondria are generally visualized as ovalshaped, they can change shape quite drastically. They swell and contract in response to various hormones and drugs. Even in very low concentrations, Lehninger found, the thyroid hormone thyroxin will make mitochondria swell and will simultaneously "uncouple" the electron transport chain from the synthesis of ATP, so that even though carbohydrates may be metabolized and oxygen consumed, no ATP will be manufactured. ATP, on the other hand, makes mitochondria contract. This swelling and contraction appear related to the movement of water in, out and through cells, which is of particular importance in the kidney.

The more scientists learn about mitochondria, however, the more their curiosity is aroused, for these are without a doubt the strangest of the cell's organelles. To begin with, mitochondria contain their own hereditary material (DNA), which resembles that of bacteria far more than that of animal cell nuclei. Because of this similarity, many scientists believe that mitochondria are derived from bacteria that infected a primitive cell and evolved into a useful symbiotic relationship with it.

As Lewis Thomas puts it, "A good case can be made for our nonexistence as entities... We are shared, rented, occupied. At the interior of our cells, driving them, providing the oxidative energy that sends us out for the improvement of each shining day, are the mitochondria and in a strict sense they are not ours.

They turn out to be little separate creatures, the colonial posterity of migrant procaryotes, probably primitive bacteria that swam into ancestral precursors of our eucaryotic cells and stayed there. Ever since, they have maintained themselves and their ways, replicating in their own fashion, privately, with their own DNA and RNA quite different from ours. They are as much symbionts as the rhizobial bacteria in the roots of beans. Without them, we would not move a muscle, drum a finger, think a thought."

"My mitochondria comprise a very large proportion of me. Looked at in this way, I could be taken for very large, motile colony of respiring bacteria, operating a complex system of nuclei, microtubules and neurons for the pleasure and sustenance of their families and running, at the moment, a typewriter," continues Thomas somewhat facetiously. "I am...obliged to do a great deal of essential work for my mitochondria.

My nuclei code out the outer membranes of each and a good many of the enzymes attached to the cristae must be synthesized by me. Each of them, by all accounts, makes only enough of its own materials to get along on and the rest must come from me. And I am the one who has to do the worrying."

During the past few years, scientists have begun to learn which proteins and enzymes of the mitochondria are synthesized according to directions in the mitochondrial DNA and which are controlled by the genes in the nucleus. This information would be of considerable importance for the design of antibiotics and drugs, as well as for the understanding of genetic diseases. For example, researchers have found that chloramphenicol–a powerful antibiotic which can cause a fatal anemia–interferes with the synthesis of mitochondrial DNA, while other antibiotics do not.

Mitochondria are just as important to the life of each cell as the heart is to the human body. At least 95 percent of our cells' energy comes from mitochondria. (Only the mammalian red blood cells, which have no nuclei, get along without mitochondria; they obtain whatever energy they need from the partial breakdown of glucose in their cytoplasm). Growing cells continuously synthesize nucleic acids, proteins and other components, for which they need the energy that is stored in molecules of ATP. Muscle cells which carry out mechanical work, plant cells which pump water against the force of gravity and many other cells also require a continuous source of energy.

Besides supplying this energy, mitochondria apparently help to control the concentration of water, calcium and other ions in the cytoplasm. They are also deeply involved in the breakdown and recycling of sugars, fatty acids and amino acids to recover their energy, as well as in the formation of urea as a waste product, to be excreted through the kidney.

These strange organelles may thus hold clues to a variety of seemingly unrelated diseases. They have been linked to some liver diseases (e.g., viral hepatitis, obstructive jaundice and cirrhosis of the liver); some muscle diseases (e.g., familial myotrophic dystrophy); and some kidney diseases. In all these conditions, pathologists have seen large amounts of materials pile up inside the mitochondria as "inclusions" of varying shapes and sizes, presumably waiting to be processed. What these inclusions consist of remains a mystery, but if scientists could find out, they might discover which specific components (for instance, enzymes) are involved in each disease.

In addition, several calcium disorders may be related to mitochondrial defects. For example, abnormal concentrations of calcium (possibly caused by malfunctioning mitochondria) may lead to arthritis, bursitis, or arterial disease. While all these leads are promising, the greatest spinoff from research on mitochondria may be in the area of energy–not only for the cell, but for general use by mankind.

As the oil crisis has made us realise, we presently depend on a limited supply of fossil fuels–coal and oil -which are of biological origin but cannot be replenished. On the other hand, nature contains a far greater resource: a molecular know-how and technology gained over millennia, by which biological systems manufacture ATP from solar energy.

This technology awaits us. It exists in very similar, if not identical forms in mitochondria and in their mirror image, the chloroplasts, which are found in green plants, where they do the work of photosynthesis, converting the sun's energy into carbohydrates. Like mitochondria, chloroplasts have an outer and an inner membrane. They have some of their own DNA. They, too, may have originated as some kind of infection, probably with blue-green algae.

And it is on chloroplasts that we ultimately depend for nearly all our energy–that of our bodies as well as that which is stored in deposits of oil and coal. If scientists ever unravel the details of the process by which chloroplasts store solar energy in carbohydrates and mitochondria release it, we may be able to produce unlimited quantities of usable energy.

MICROTUBULES, THE CELL'S PHYSICAL PROPS

All the organelles described so far are bound by membranes. For a long time, the rest of the cytoplasm–a liquid called the cell sap, or ground substance–appeared totally unstructured. But as scientists learned to use newer and gentler fixatives to prepare cells for electron miscroscopy, more and more tiny structures materialized in this soup.

The most prominent of the new structures are two organelles which provide an intricate system of physical support for the cell–the equivalent of buttresses or skeletal bones–as well as a contractile mechanism for cell movement: microtubules and microfilaments. Neither is bound by a membrane.

Microtubules were first noticed in the middle 1950's, but they were seen only rarely until the development of glutaraldehyde as a gentle fixative in 1963. Extremely thin cylinders, about 200 to 300 angstroms in diameter and of variable length, microtubules are constructed chiefly of proteins called tubulins. In 1967, Edwin Taylor and his associates at the University of Chicago discovered that colchicine–a chemical used to arrest cell division–did its work by binding to the protein tubulin. This pointed to the role of microtubules in mitosis and helped to answer a question which had worried generations of biologists: How do chromosomes sort themselves out into two sets and separate during cell division? The details of mitosis would have been very hard to explain without understanding the chemistry of the spindle fibers which organize this separation.

In fact, if there were no microtubules, scientists would have had to invent them. Something of this sort was needed to account not only for mitosis, but for the surprisingly firm structure of such seemingly vulnerable fibers as the long, thin axons of nerve cells, as well as for the unique structure of some red blood cells, which are held in a disc-like shape by hoops composed of microtubules.

Microtubules are also involved in assisting the transport of substances in and out of the cell and in the motion of cells themselves. They make up the cilia and flagella, whip-like filaments which project from the surface of cells and move rhythmically. Large numbers of cilia are found on cells that line the respiratory system, for instance, where they help to sweep out dust and debris. Both cilia and flagella play an important role in human reproduction: the coordinated beating of cilia in the oviduct produces a sort of current which draws the female's egg into the uterus, while the rapidly thrashing tail of sperm actually is a flagellum.

MICROFILAMENTS

The movement of living cells has fascinated biologists ever since Leeuwenhoek discovered "tiny animalcules prettily a-moving" under his simple microscope. But for a long time, scientists' efforts to study such movements were stymied by techniques which required killing cells before they could be examined under the microscope. It is only since the development of time-lapse photography, together with the phase-contrast and interference microscopes, that researchers have been able to study cells while they were moving.

There seem to be two main kinds of cell motion: That produced by the rhythmic beating of special structures on the cell surface (such as cilia and flagella) and that connected with some wispy organelles inside the cell, microfilaments. The action of these microfilaments allows cells to "crawl" along surfaces by forming extensions, called pseudopods, towards which the bulk of the cytoplasm flows.

While the precise nature of this movement is not understood, it seems to involve the continual transformation of the liquid parts of the cytoplasm (the cell sap, or ground substance) into a viscous gel with the help of calcium and its subsequent dissolution into a liquid again. Amoebae, white blood cells, macrophages and the tips of growing nerve cells "crawl" in this fashion. So, probably, do cancer cells.

Microfilaments apparently consist of thin strands of actin, a protein. Actin has been studied for nearly half a century in muscle cells. In the fifties scientists learned, primarily through the work of Hugh Huxley at the Medical Research Council laboratories in England, that contraction in the skeletal muscles is produced by the sliding of thin filaments of actin between thick filaments of another protein, myosin.

But it is only in the past 15 years that researchers have found evidence of similar actin-myosin interactions, involving microfilaments, in many other kinds of cells. The whole process of secretion, as when cells discharge hormones, enzymes or other proteins, may depend on such interaction.

So may the movement of many cells which help to defend the body against invading organisms. As the President's Biomedical Research Panel put it recently, "Here again is a striking example of biological versatility and with it the promise of a common key to the solution of many problems in basic or applied biomedical sciences." Current research on all the cell's organelles–studies of the nucleus, ribosomes, ER, Golgi, lysosomes, mitochondria, microtubules and microfilaments holds such promise. However the common key that may unlock the greatest number of health benefits may well be found in the cell's filmy membranes, particularly in the plasma membrane at the surface of the cell.

MEMBRANES

Whatever flash of lightning there was that organized purines, pyrimidines and amino acids into macromolecules capable of reproducing themselves, it would not have yielded cells but for the organizational trick afforded by the design of a membrane wrapping." He visualizes a kind of bubble in which the first macromolecules must have been enclosed, to protect them from being dissipated or dissolved in the strong salt of the early sea.

Membranes are now arousing enormous curiosity among biologists because it has become clear that these diaphanous films, less than 100 angstroms (1/100,000th of a millimeter) thick, control some of the most vital functions of the living cell. Furthermore, there are layers upon layers of them, in and around various organelles.

The double-membrane envelope which forms a channel around the nucleus not only protects the genetic material but also connects with the "plumbing" the system of membranous tubes, folds and compartments -which makes up the rough and smooth endoplasmic reticulum and the Golgi apparatus. The membranecovered mitochondria contain a separate inner membrane system of their own. And at the surface of the cell, the all-powerful plasma membrane–an organelle in its own right–stands between the cell and the world.

Like the skin that covers our bodies, the plasma membrane creates a separate environment within which the biochemical processes of life can take place. In addition, it controls everything that goes in or out of the cell. It also receives signals from other cells, interacts with them to form tissue, recognizes attacking viruses, responds to hormones. And as a result of its contact with the outer environment, it probably tells the nucleus when to divide–or when to stop. Despite its fragility, it is a tough and resourceful guardian of the cell's interior.

The secret of the membranes' power lies in their ingenious composition and structure. After years of puzzlement and speculation, biologists now have some idea of how these thin membranes are organized. Of course there is no "typical" membrane, since the various types that exist within each cell and in different kinds of cells differ in composition and function. Furthermore, any particular membrane is likely to change from moment to moment, in response to its environment. Nevertheless all cell membranes follow the same general pattern.

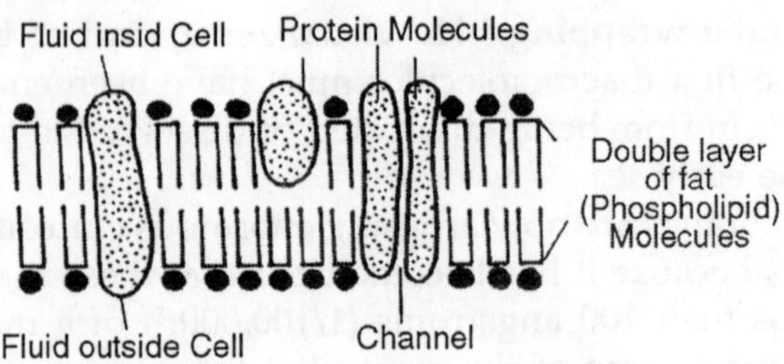

Fig. The Plasma Membrane

Their first task is to provide effective compartments. Since cells live in a watery environment and also contain a large amount of watery fluid, in which the organelles float, anything which forms a compartment must be pretty water tight. In fact, membranes have been found to consist mostly of lipids (fat) interspersed with proteins and since fat and water don't mix, this creates an efficient barrier to keep foreign substances out and water in. Substances that dissolve well in lipids will not dissolve well in water and vice versa (excepting only detergents and gases, such as oxygen and nitrogen, which dissolve in both).

A membrane is not just a layer of fat, however, but an intricate, elastic and fluid "bilayer" consisting of two layers of phospholipids (each layer one molecule thick) interspersed with proteins which are either embedded in the lipid layers or loosely attached to their surface. The two layers of phospholipids arrange themselves very neatly, as described in 1938 by James Danielli, who was then at the State University of New York at Buffalo: The part of the phospholipid molecule which is water-loving (a small portion containing charged phosphate, which is electrostatically attracted to water) faces out towards the outside world, or towards the watery cytoplasm, while the rest of the molecule (consisting of hydrocarbons, which are insoluble in water) is pushed to the centre of the membrane. Actually the layer of membrane which faces the outside world fulfills quite different functions from the layer which faces the cytoplasm. Therefore the two layers are not symmetrical; each consists of a different array of lipids and proteins.

MEMBRANE PROTEINS

In 1966, S.J. Singer of the University of California at San Diego proposed a "fluid-mosaic" model for cell membranes. He compared some of the proteins in cell membranes to "icebergs floating in a sea of lipids," and said that their movement would create an ever-changing "fluid-mosaic" pattern. In his view, most of these proteins are folded up so as to have one hydrophobic (waterhating) and one hydrophilic (water-loving) end–just as do the phospholipid molecules, though of course the proteins are much larger.

He also suggested that if the proteins are large enough to span the entire thickness of the membrane, they must have two hydrophilic ends (which protrude) and a hydrophobic middle (which is embedded in the phospholipid layer). Although such tri-partite proteins were unknown at the time, they have been found to exist in membranes.

The proteins in membranes are extremely important because they do most of the specialized work, while the lipids serve largely as a diffusion barrier. In the plasma membrane, for example, protein "receptors" respond to specific hormones, neurotransmitters and antibodies. They selectively accept specific chemicals from outside the cell, which must fit them as precisely as a key in a lock and send appropriate signals into the cell's interior.

The receptors on which so much depends must make exquisitely refined discriminations at every phase of the cell's life, beginning with fertilization. For instance, sperm cells recognize egg cells only when these belong to the same or to a very closely related species. As George Palade puts it, "Imagine what problems could arise for all the creatures that live in the sea, with all kinds of eggs and all kinds of sperm floating around in the water! Yet because of the precise information obtained through receptors on the cell membranes, fertilization occurs only within specific species." The intricate process of differentiation, through which embryonic cells take on specialized functions, also depends largely on specific cells' recognizing one another through the proteins in their plasma membranes.

The failure of protein receptors to do their job properly may lead to a variety of diseases. "We are beginning to understand that there may be either inherited or acquired defects in the organization of cell membranes," says Palade, who now directs the Centre for Research on Cellular Membranes which NIGMS started at Yale University in 1974. "For example, an organism may produce enough of a hormone, such as parathormone or insulin, but not have enough receptors for it in the membrane of the target cell.

This may look like a hormone deficiency, when in fact it is not. We are finding many membrane disturbances of this sort. Or there may be an excess of receptors on the membrane of the target cell, leading to a syndrome that mimics the overproduction of a hormone or neurotransmitter; for example, some forms of hypertension and cardiovascular disease may be produced by an excess of receptors for hormones on the membranes of smooth-muscle cells on the arterial walls."

PLASMA MEMBRANE

As primary guardian of the cell's gates, the plasma membrane is responsible for keeping the chemical composition of the cell's interior constant within very narrow limits. The interior of the cells of higher animals needs to be high in potassium ions and low in sodium ions, for instance, while their exterior environment–blood and body fluids–has the reverse concentration, being high in sodium (like sea water) and low in potassium. This creates a difference in electrical potential of about one-tenth of a volt or higher across the cell membrane, which is particularly important to the action of nerves and muscles. Scientists are only beginning to understand how the plasma membrane accomplishes such impressive feats. The simplest traffic across the plasma membrane is passive transport, which requires no energy. The molecules diffuse through the membrane at rates determined by their solubility in lipids, their size and differences in their concentration. Those that are highly soluble in lipids move most rapidly. Water and other small molecules which are not soluble in lipids can also diffuse through the membrane, with the smallest ones passing through at the highest rate.

(If there were no membrane at all, however, water could diffuse in or out of the cell 100,000 times more rapidly. It is believed that there are "channels" or "pores" across the lipid layer through which water can move freely; if so, the total area of these channels must be 100,000 times smaller than the total area of the cell surface). In each case of passive transport, the molecules diffuse from a space where their concentration is high to a space where their concentration is low, until it is equal on both sides of the membrane.

With sodium and potassium, however, some mechanism actually moves ions in and out of the cell against the concentration "gradients." This active transport, which maintains a high concentration of potassium and a low concentration of sodium within the cell, despite opposite conditions on the other side of the membrane, requires the expenditure of energy. And the energy comes from the cell's universal fuel, ATP.

The plasma membrane contains certain enzymes which split the chemical bonds of ATP and then use the energy to carry potassium in and sodium out of the cell. The amount of energy involved is surprisingly high: It has been calculated that up to 18 percent of the total ATP required by the brain, for instance, is expended in such activities.

A third type of traffic across the plasma membrane involves the release or uptake of various substances whose molecules are too big to cross the membrane either by diffusion or through pores. Instead, they travel in membranous packages. In protein-secreting cells, for instance, new proteins are wrapped in pieces of membrane in the Golgi apparatus, forming vesicles; the vesicles then move to the cell's surface, where their membranous envelope fuses with the plasma membrane and their contents are dumped outside the cell.

This is called exocytosis. In a reverse process, endocytosis, proteins or other molecules from outside the cell attach to the cell's plasma membrane, which folds in to envelop them; this forms small vesicles which move into the cytoplasm, carrying drops of extra-cellular medium with the substance to be used inside the cell. In all of these cases, matter is brought into the cell or discharged from the cell without interrupting the continuity of the plasma membrane. Something similar occurs during fertilization, when the sperm's plasma membrane fuses with the egg's plasma membrane in such a way that the sperm nucleus can enter the egg without any exposure to the surrounding medium.

Besides directing traffic in and out of the cell in all these ways, the plasma membrane controls the cell's communications system, receiving signals from the outer environment (including signals from other cells) and sending out messages of its own. Some of these signals are carried by the chemical messengers called hormones, which circulate in the bloodstream with instructions to cells in many parts of the body. As these hormones reach each target cell, they generally do not enter it, but trigger intricate responses.

Only a few hormones, such as the steroids (including cortisone and the sex hormones) which are derived from cholesterol and thus easily soluble in fat, can actually pass through the cells' membranes. Most hormones, for instance insulin and some neurotransmitters, act by giving signals to specific receptors on the cell surface; these receptors then activate a second-messenger system which orders appropriate changes within the cell.

Earl Sutherland of Vanderbilt University discovered how this second-messenger system works and won the Nobel Prize for it. He found that adenylate cyclase, an enzyme in the plasma membrane of many cells, can be activated to convert ATP (the universal currency of energy in cells) into cyclic AMP, a nucleotide which is then released into the interior of the cell. Thus, when a hormone (the first messenger) binds to a specific receptor on a cell surface, it triggers the dispatch of cyclic AMP (the second messenger) to the cell's biochemical machinery. Another cyclic nucleotide which acts in similar fashion, cyclic GMP, has been identified and recently scientists discovered yet a third type, cyclic CMP.

It is now clear that these second messengers regulate hundreds of diverse processes in living cells. The surge of energy which we sometimes feel under stress (as a result of the conversion of stored glycogen into glucose); the discharge of secretory products from cells; the activation of dormant genes; and even contact inhibition–all depend on the action of a second messenger on different components of different cells. The second-messenger system is involved in so many diseases, in fact, that substances which facilitate or inhibit it are expected to prove very useful in medicine in the future.

Hormonal messages travel far and wide in the body via the bloodstream; they can be recognized by very distant organs, but they may take hours, days or weeks to produce their complex effects. In order to stay alive, animals must have a much quicker means of response, with which they can react to events within seconds or milliseconds. This is where the nervous system comes in. When you touch a hot stove, look or listen, the information is carried almost instantaneously by neurons (nerve cells) which, unlike most other cells, produce electrical signals in response to physical or chemical stimuli. The neurons' plasma membranes play a key role in this transmission.

One common sequence is as follows: An electrical signal travels down the surface of a neuron's thin axon (a long fibre); this induces vesicles which contain units of neurotransmitter to fuse with the plasma membrane at the terminal of this axon and to release their contents into the synaptic cleft (the gap between this neuron and the next cell). If enough of the newly released transmitter reaches the plasma membrane of the next neuron, it changes the membrane's electrical potential, producing a new nerve impulse which then repeats the whole process. If, on the other hand, enough transmitter reaches the plasma membrane of a muscle cell at a special region called the neuromuscular junction, it stimulates the muscle to contract.

NEW TOOLS WHICH OFFER A NEW VIEW OF MEMBRANES

As important and powerful as the plasma membrane is, it makes up only a fraction of the total amount of membrane in a cell. Most animal cells, especially those that have specialized functions, are so packed with membranes that there may be 50 to 100 times as much membrane inside them as on their surface. (In plant cells the proportion of inner membrane is even greater, going up to 150 times the amount of membrane on their surface). Yet the plasma membrane is best known to scientists. The main problem in studying membranes has been the difficulty of obtaining pure specimens. Until recently, this was possible only with the plasma membrane of the mammalian red blood cell, which has no nucleus.

When biochemists grind up cells and centrifuge their components, only fragments of the cells' membranes can be recovered–usually in the form of small, closed vesicles or ragged pieces of unidentified origin. These can be sorted out and identified, so that by now the membranes of practically every organelle can be isolated, but they still vary in purity. Such brutal treatment is not necessary with the mammalian red blood cell, however.

Since it contains mostly hemoglobin and has neither nucleus nor mitochondria, nor any internal membranes, it can be emptied with relative ease: It is simply induced to swell up until temporary, large pores open in its plasma membrane, letting the fluid and hemoglobin seep out. The plasma membrane is then allowed to shrink back to its normal size and the pores close, leaving researchers with an intact outer membrane or "ghost." Although other kinds of membranes can now be isolated by various means, the mammalian red blood cell "ghost" remains the favourite subject of many scientists who work on membranes.

Studying these membranes under an electron microscope proved frustrating, however. Since electrons can pass only through ultra-thin sections, the cell was usually sliced finely with a straight knife to prepare samples for the microscope. Yet cell membranes are far from straight and it was practically impossible to get a fullface view of a membrane in this fashion. Nor could one learn much about the membrane's inner layer.

In the early 1960's a Swiss scientist, Hans Moor, developed a method which offered an entirely new perspective on cells. Instead of chemically fixing and staining the cells, he froze them rapidly, following a technique which had previously been used in California with viruses.

Then, instead of cutting the cell tissue into smooth sections, he fractured it along natural lines of weakness and bombarded the exposed surfaces with vaporized carbon and platinum, which condensed and hardened on these frozen surfaces. After he allowed the specimens to thaw out, there remained a very thin and highly detailed metal replica of the fractured plane–a replica thin enough to be viewed under an electron microscope.

Moor naturally assumed that he was looking at replicas of the cells' outer surfaces. But in fact, as Daniel Branton, who was then at the University of California, Berkeley, suggested in 1963, an extraordinary thing was happening: the membranes were being cleaved down the middle in two separate layers. "This seemed a very outrageous idea at the time and no one believed it," recalls Branton, who is now at Harvard University. Yet there was a logical explanation for it: The two lipid layers in the centre of the membrane were only weakly held together by hydrophobic interactions and when the water froze, these interactions became irrelevant, so the membrane fell apart.

By 1967 most researchers were convinced. As they began using the freeze-fracture technique to explore the interior of membranes, they discovered little balls–protein particles–some of which were invisible from the outer surface. Furthermore, they found that the outer half of the membrane sometimes differed quite radically from the inner, cytoplasmic half, both in protein structure and in lipids. This enabled them to zero in, for the first time, on specific proteins and lipids in the centre of the membrane.

While the freeze-fracture technique opened up the interior of the membrane for study, the true surface of cell membranes became visible through a special kind of electron microscope, the scanning electron microscope (SEM), which provides three-dimensional images. Unlike the transmission electron microscope, whose beams pass through the specimen, the SEM depends on low-energy electron beams which scan the specimen's surface, exciting secondary electrons whose pattern produces the image. Until recently, the SEM could not resolve more than 150 angstroms. "The pictures were gorgeous, but there was very little information coming out of them," recalls Palade.

However, a new instrument developed in Japan is presently making it possible to see down to 30 angstroms. "So now we can start seeing the molecular details of the cell surfaces," he says. "We know that receptors in the plasma membrane must protrude from the surface of cells, but we don't know the size of these protrusions.

We want to be able to obtain information about them, not only by biochemical techniques. We want to know the frequency and density of these molecules, their pattern of distribution on the cell surface, whether this distribution can be disturbed and whether any disease states are connected with defects in it. This is the excitment about the high-resolution SEM: It brings you down to a level at which you can really look at molecules!"

To make full use of this instrument's power, the procedures for specimen preparation must also be refined by a factor of five. This requires new methods of "fixing" the specimens with chemicals and also new ways of spraying their surface with metal vapour (to make them give off secondary electrons).

One advantage of working in an interdisciplinary centre such as the Yale Centre for Research on Cellular Membranes is that it is large enough to maintain experts who can devise such techniques. The Centre consists of seven different research groups, each with its specific programme and each depending, at least in part, on advanced technology in two areas: protein chemistry and the SEM. It includes some twenty-five cell biologists, biochemists, pathologists and biophysicists. "None of us uses the SEM or protein chemistry facilities full time," Palade explains, "and none of us could run these facilities with an adequate amount of attention and competence by ourselves. Here we can all use them profitably."

The Center's protein chemistry lab routinely analyzes extremely small samples of protein extracted from cell membranes. It boasts a very effective amino-acid analyzer, controlled by a computer, which uses highpressure liquid chromatography to identify the different amino acids in a protein quickly and accurately. It also has a sophisticated machine which can determine the sequence of these amino acids from a sample of 50 micrograms or even less. Soon the Centre hopes to have an additional instrument to assess the radioactivity in labelled amino-acid residues–an important tool with which to study the origins and fate of membrane proteins.

Palade's own group of 10 scientists concentrates on unraveling the "functional interaction" between cell membranes and on the membranes' genesis. "The membranes that form compartments within cells are usually separate from one another, but some of them fuse, establishing continuity," Palade says. "This is important for secretory processes, when a cell product moves from one compartment to another and then is discharged. The membranes must know the partner with which to fuse; how do they communicate? Every time a cell divides, too, the plasma membrane fuses to split the cell and then the two daughter cells separate.

We are trying to understand this fusion-fission process. At present we only understand the recognition of a hormone or a neurotransmitter by a receptor on the plasma membrane. But we believe the same kind of thing happens inside the cell when membranes fuse, except that both molecules -the key and the lock–are bound to membranes. The basic principle is the same: recognition by complementarity."

Where do all these membranes come from? As Palade points out, "The cell has discovered a way to increase its membranes without disturbing their function. It appears to use always its pre-existing membranes, which behave like a fine film of olive oil and puts new lipid and protein molecules into it, so that the film expands without ever breaking its continuity. Cells never begin making a membrane from scratch. When they divide, they give their membranes in equal parts to their daughter cells. In fact, we inherit our membranes from our mothers.

"All these membranes have their components replaced continually," he adds. "So there is a continuity of pattern, though not of substance. What we want to find out is where the components–the proteins and the lipids–are produced and how they are assembled. We'd particularly like to know where the glycoproteins (sugarbearing proteins), some of which seem to be the receptors of the plasma membrane, are synthesized.

"It's partly a matter of scientific curiosity, but there are also practical reasons: When cells have defective receptors, the actual defect may be inside the cell, rather than on the surface. Maybe the components are not produced in sufficient quantity. Maybe they are produced but are not transported or assembled properly. This may be relevant to cancer, as well as to questions of growth and of aging."

MEMBRANE PROTEINS

One of the Yale Center's most exciting findings so far has concerned the role of "glycophorin," a sugar-bearing protein that spans the entire thickness of the red blood cell membrane. Named by Vincent Marchesi, who succeeded in detaching it, intact, from membrane ghosts, glycophorin has some carbohydrate chains which protrude from the membrane surface like antennae and serve important functions: They act as the cell's recognition sites for blood groups and for certain viruses. In addition, they are involved in maintaining the negative charge on the cell's surface which prevents it from clumping with other red blood cells.

Glycophorin is now the best-known of the glycoproteins that extend through the cell membrane and the first to have its amino acids completely analyzed and sequenced. Immense quantities of blood were needed to achieve this knowledge.

(Twelve gallons of human blood can provide only 125 grams of membrane ghosts, which produce only 3 grams of pure protein for analysis).

Just recently, Marchesi's research group made the further discovery that under certain conditions glycophorin exists in twin form, with a narrow space to separate the twin shapes, creating an open channel all the way through the membrane. Marchesi believes that these open channels are the long-sought "pores" through which materials can flow rapidly across a membrane. He also believes that glycophorin has brought him to the verge of answering several important and interrelated questions about health and disease, all of which involve the mechanism of cellcell recognition.

For example, there is the mystery of blood platelets - billions of little bodies that float about in blood and pile up at cuts in the blood vessels to prevent the vessels from leaking. Normally these platelets do not stick to one another, but somehow, when they sense an injury to the blood vessel, their surface membranes change and they begin to clump. The mechanism involved here is very similar to that of normal cell-cell recognition, Marchesi points out. Yet in certain cases it may aggravate arteriosclerosis and other forms of cardiovascular disease.

Then there is the lure of understanding the real function of carcinoembryonic antigen (CeA), a protein found on the plasma membrane of some human tumors as well as on normal embyronic tissue. Surprisingly, CeA is an almost exact duplicate of glycophorin, differing only subtly in composition and immunological reactivity. Yet it can be used to measure the progress of certain kinds of cancer therapy: After an operation on colon cancer, for example, the level of CeA in the blood drops to almost zero, going up again if new tumors begin to grow. Researchers would like to develop other CeA-based tests which might reveal the presence of cancer very early, when it is easiest and safest to treat. This might be done through annual blood tests. However, they are hampered by "the lack of a deep enough understanding of what's going on in normal membranes," says Marchesi.

Current methods of analyzing membrane proteins are "much too simple, because what we can do is only twodimensional," Marchesi explains. "We can see only the gross differences between molecules. But there are very subtle ways in which these little chains of peptides can fold over each other. That's the exciting part! We need to develop methods for studying these membrane proteins in three dimensions. There's an infinite amount of complex communication going on between cells, yet probably only a few molecules mediate it, like a translation system or a code. So that's the trick–to find the code. As you know, the DNA code is only four letters, in a linear arrangement. This code, we think, is hooked up in the 3-D arrangement of the molecules. It's probably going to be a lot more subtle than the code for DNA."

New techniques are being developed towards this goal by Lubert Stryer, formerly of the Yale centre and now at Stanford Medical School. The only proteins whose 3-D structure has been worked out by scientists so far are some 20 water-soluble proteins–for example, hemoglobin, Stryer says. Analyzing the 3-D structure of insoluble proteins, such as those in membranes, will pose a far greater challenge, he notes, since at present these proteins cannot be crystallized. Researchers in Stryer's lab are starting out with various techniques taken from physics. They are doing optical rotation experiments to see whether glycophorin has any regular structures such as a helix. They are using a Raman spectrometer, shining laser light of a known wave length onto a glycophorin sample and studying the pattern of light that it scatters as a clue to its molecular structure. They are inserting fluorescein-labelled glycophorin into an artificial membrane, to find out whether it acts as a channel for sodium and potassium and whether altering the part of the molecule that protrudes on one side of the membrane will affect the part on the other side. They are also collaborating with the Brookhaven Laboratories in neutron studies of the molecular structure of rhoclopsin, a membrane protein that is involved in vision.

"All this technology is very, very recent," says Stryer, "but it may be extremely powerful. We are building up an image of how rhodopsin sits in this membrane–it may actually go through the membrane. If rhodopsin turns out to be a gate that is opened by light, other gates in membranes may be opened by hormones in a similar way. That's our hope."

THE PROMISE OF NEW THERAPIES

Throughout the world, researchers who work on cell membranes realise that they have only just begun to explore a field in which anything seems possible. They share a great excitement about it.

"Can we expect to do 'membrane engineering' in the future?" S.J. Singer asked a scientific gathering recently. And he answered with a resounding, "Yes!" For example, it may be possible to change the mobility of membrane components, he says. This might have extraordinary results. Normal cells have some kind of matrix underlying their membranes which restricts the mobility of certain membrane components, he points out. Yet when cells are malignantly transformed, their membrane components can suddenly move about more easily. This on-off control of the components' mobility may well be the mechanism of growth, Singer believes; the oily film of lipids in the membrane may need to become more fluid to allow for the insertion of new molecules as the membrane grows. The same on-off control might also be a key to cancer. Scientists might then restore malignant cells to normality by providing anchoring points for the components of their plasma membranes.

Physicians already do modify the plasma membrane of their patients' cells when they use various general anesthetics, which dissolve in membrane lipids and change their properties. In the future, it is almost certain that drugs will be tailor-made to act more selectively either on proteins or on lipids in the plasma membranes of specific kinds of cells.

Some of these drugs may themselves be enclosed in artificial membranes, forming "liposomes." In this way they could be engineered for release only when or where needed, by means of specific receptor molecules which would be inserted in the liposome membrane. "For example," suggests

Cambridge University's Alec Bangham, "liposomes containing a tranquilizer could be equipped with molecules that only respond to supernormal levels of adrenalin." Similarly, liposomes loaded with antitumor agents could be keyed to the properties of specific tumor cells so that the drug would be released only when the liposome came in actual contact with the malignant cell. "By this method," Bangham explains, "It would be possible to employ drugs far too toxic for administration through the general circulation." Experiments of this sort are already under way in Christian de Duve's lab and in other places.

Other researchers have been working on ways to correct the intricate feedback system through which cells control the production of such vital chemicals as cholesterol, in cases when the system fails. As Joseph Goldstein and Michael Brown of the University of Texas discovered recently, each person's production of cholesterol is controlled by the activity of specific receptors for LDL (low-density lipoproteins, the particles which carry cholesterol in the blood) in his or her cells' plasma membranes–and these receptors in turn increase or decrease according to the amount of cholesterol in the cell.

Mammalian cells can both produce their own cholesterol and take up cholesterol from the blood. When a normal cell needs more cholesterol than it has produced, it synthesizes more LDL receptors. These transmit LDL from the blood to the cytoplasm, where lysosomes break it down, releasing cholesterol. As soon as the cholesterol in the cytoplasm reaches a certain concentration, however, it shuts off the enzyme that produces cholesterol inside the cell–and also stops the synthesis of additional LDL receptors.

The system works pretty much like a thermostat to ensure that the cell always has enough cholesterol for life and growth, without accumulating too much of it. Because of a defective gene, however, some people do not have a sufficient number of LDL receptors in their cells. This is what happens in familial hypercholesterolemia, where the lack of LDL receptors breaks the normal chain of control and leads to overproduction of cholesterol. Among people who suffer heart attacks before the age of 60, one out of 20 have been found to carry this defective gene (which actually occurs in about one out of 500 persons in the general population).

Now that scientists understand the feedback system involved, they realise why existing anticholesterol drugs are so often ineffective: lowering the amount of cholesterol in the blood may actually lead the flawed cells to produce even more of it, to make up for the loss. By contrast, new drugs which are now being tested would specifically reduce the cell's ability to make cholesterol, rather than remove more of the cholesterol that is already made.

Many forms of mental illness may also turn out to be related to defects in specific proteins that act as receptors for hormones and neurotransmitters on the plasma membrane. If so, it may be possible to develop far more effective drugs which would take account of the cell's feedback systems. Every new bit of information about the inner workings of the cell leads towards the development of better weapons against the diseases that still plague us. As researchers learn more about membranes, for example, they are indirectly contributing to the prevention or cure of such seemingly unrelated ailments as diabetes, heart disease, various genetic defects, cancer and even kidney disease, for the kidney is composed largely of membranes. New vaccines and new immunologic weapons also await a better understanding of the cell and its organelles.

While today's physicians view diseases in terms of organs, such as liver diseases, bone diseases or heart diseases, the physicians of the next generation will see these ailments as diseases of cell membranes and organelles -for example, diseases of the endoplasmic reticulum, lysosomal diseases, or mitochondrial diseases. Thus, heart disease will be examined not simply in terms of which chamber and portion of the heart is involved, as determined by the EKG, but in terms of the specific lesion within the heart muscle cells that leads to malfunction and cell death.

This has been a period of "spectacular growth" in cell biology, the President's Biomedical Research Panel declared recently. For the first time in the history of the life sciences, the essential function of cells can be understood in terms of defined chemical reactions." Yet despite all this progress, most of medicine is still far from being a true science.

The reason for this gap is that, until now, scientists could not even study the mechanisms which regulate such basic processes as cell growth, division, or mobility; they simply did not have enough information about the cell's components. They are barely starting to understand how cells are controlled. But now the stage is set. A far more powerful type of research on the cellular and molecular basis of disease can begin. And as it builds up a clearer picture of normal and abnormal cell function, it will create a truly scientific medicine for the 21st century.

Chapter 4

Evaluation of Bacteria

Bacteria are ubiquitous in every habitat on Earth, growing in soil, acidic hot springs, radioactive waste, water, and deep in the Earth's crust, as well as in organic matter and the live bodies of plants and animals. The Bacteria are a large group of unicellular microorganisms. Typically a few micrometres in length, bacteria have a wide range of shapes, ranging from spheres to rods and spirals. There are typically 40 million bacterial cells in a gram of soil and a million bacterial cells in a millilitre of fresh water; in all, there are approximately five nonillion (5×10^{30}) bacteria on Earth, forming much of the world's biomass.

Bacteria are vital in recycling nutrients, with many important steps in nutrient cycles depending on these organisms, such as the fixation of nitrogen from the atmosphere and putrefaction. However, most bacteria have not been characterized, and only about half of the phyla of bacteria have species that can be cultured in the laboratory. The study of bacteria is known as bacteriology, a branch of microbiology.

There are approximately ten times as many bacterial cells as human cells in the human body, with large numbers of bacteria on the skin and in the digestive tract. Although the vast majority of these bacteria are rendered harmless by the protective effects of the immune system, and a few are beneficial, some are pathogenic bacteria and cause infectious diseases, including cholera, syphilis, anthrax, leprosy and bubonic plague. The most common fatal bacterial diseases are respiratory infections, with tuberculosis alone killing about 2 million people a year, mostly in sub-Saharan Africa.

In developed countries, antibiotics are used to treat bacterial infections and in various agricultural processes, so antibiotic resistance is becoming common. In industry, bacteria are important in processes such as sewage treatment, the production of cheese and yoghurt through fermentation, as well as biotechnology, and the manufacture of antibiotics and other chemicals.

The Bacteria are a group of single-cell microorganisms with procaryotic cellular configuration. The genetic material (DNA) of procaryotic cells exists unbound in the cytoplasm of the cells. There is no nuclear membrane, which is the definitive characteristic of eucaryotic cells such as those that make up plants and animals. Until recently, bacteria were the only known type of procaryotic cell, and the discipline of biology related to their study is called bacteriology. In the 1980's, with the outbreak of molecular techniques applied to phylogeny of life, another group of procaryotes was defined and informally named "archaebacteria".

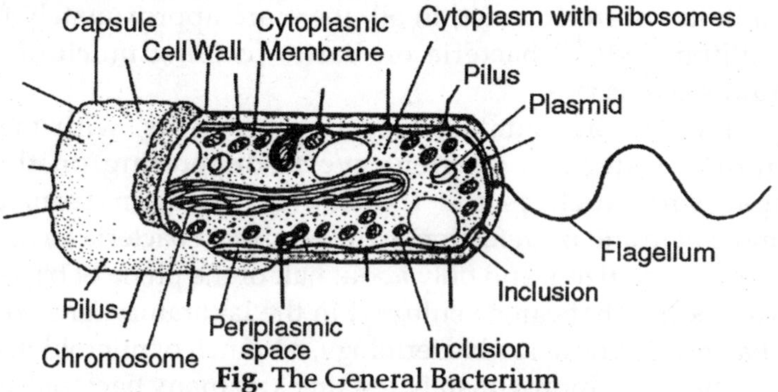

Fig. The General Bacterium

This group of procaryotes has since been renamed Archaea and has been awarded biological Domain status on the level with Bacteria and Eucarya. The current science of bacteriology includes the study of both Domains of procaryotic cells, but the name "bacteriology" is not likely to change to reflect the inclusion of archaea in the discipline. Actually, many archaea have been studied as intensively and as long as their bacterial counterparts, except with the notion that they were bacteria.

EVALUATION OF LIFE ON THE EARTH

4 billion years ago when life arose on Earth, the first types of cells to evolve were procaryotic cells. For approximately 2 billion years, procaryotic-type cells were the only form of life on Earth. The oldest known sedimentary rocks, from Greenland, are about 3.8 billion years old. The oldest known fossils are procaryotic cells, 3.5 billion years in age, found in Western Australia and South Africa. The nature of these fossils, and the chemical composition of the rocks in which they are found, indicate that lithotrophic and fermentative modes of metabolism were the first to evolve in early procaryotes. Photosynthesis developed in bacteria a bit later, at least 3 billion years ago.

Anoxygenic photosynthesis (bacterial photosynthesis, which is anaerobic and does not produce O_2) preceded oxygenic photosynthesis (plant-type photosynthesis, which yields O_2). However, oxygenic photosynthesis also arose in procaryotes, specifically in the cyanobacteria, which existed millions of years before the evolution of green algae and plants. Larger, more complicated eucaryotic cells did not appear until much later, between 1.5 and 2 billion years ago.

The archaea and bacteria differ fundamentally in their structure from eucaryotic cells, which always contain a membrane-enclosed nucleus, multiple chromosomes, and various other membranous organelles such as mitochondria, chloroplasts, the golgi apparatus, vacuoles, etc. Unlike plants and animals, archaea and bacteria are unicellular organisms that do not develop or differentiate into multicellular forms. Some bacteria grow in filaments or masses of cells, but each cell in the colony is identical and capable of independent existence. The cells may be adjacent to one another because they did not separate after cell division or because they remained enclosed in a common sheath or slime secreted by the cells, but typically there is no continuity or communication between the cells.

There are usually exceptions to the lack of rules in biology. In the case of communication among bacterial cells, we now understand how certain bacteria in culture and in nature are able to communicate with one another through a process called quorum sensing. For example, a bacterial culture may have to first produce a critical concentration of a chemical, and therefore reach a critical density of cells, in order to perform some function such as turn on light production (bioluminescence) or toxin production, or to begin cell division or to engage in certain processes of DNA exchange.

THE UNIVERSAL TREE OF LIFE

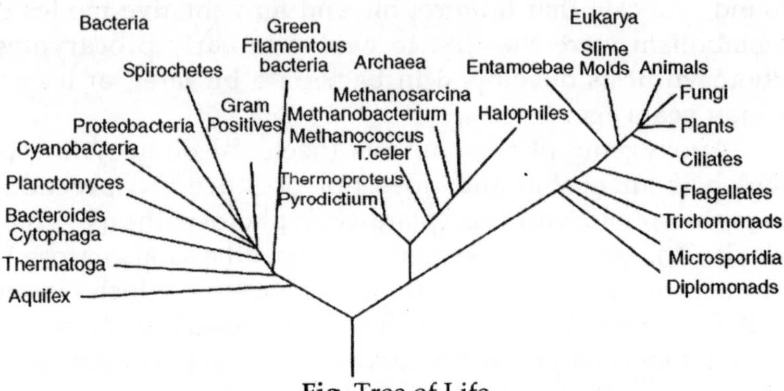

Fig. Tree of Life

On the basis of small subunit ribosomal RNA (ssrRNA) analysis the Woesean Tree of Life gives rise to three cellular "Domains": Archaea, Bacteria, and Eucarya. Bacteria (formerly known as eubacteria) and Archaea (formerly called archaebacteria) share the procaryotic type of cellular configuration, but otherwise are not related to one another any more closely than they are to the eucaryotic domain, Eucarya. Between the two procaryotes, Archaea are apparently more closely related to Eucarya than are the Bacteria. Eucarya consists of all eucaryotic cell-types, including protista, fungi, plants and animals.

PROKARYOTIC EVOLUTION

Comparison of protein sequences whose genes are common to the genomes of several procaryotes has resulted in a "genomic time scale of procaryotic evolution" and establishes the following dates for some major events in procaryotic evolution. The results are consistent with most other phylogenetic schemes that recognize higher-level groupings of procaryotes.

The time estimates for methanogenesis support the consideration of methane, in addition to carbon dioxide, as a greenhouse gas responsible for the early warming of the Earths' surface. Divergence times for the origin of anaerobic methanotrophy are compatible with carbon isotopic values found in rocks dated 2.8 - 2.6 Ga.

The origin of phototrophy is consistent with the earliest bacterial mats and structures identified as stromatolites, but a 2.6 Ga origin of cyanobacteria suggests that those structures (if biologically produced) would have been made by anoxygenic photosynthesizers. A well-supported group of three major lineages of Bacteria (Actinobacteria, Deinococcus, and Cyanobacteria), that have ben called "Terrabacteria", are associated with an early colonization of land.

BACTERIA AND ARCHAEA

Most procaryotic cells are very small compared to eucaryotic cells. A typical bacterial cell is about 1 micrometer in diameter or width, while most eucaryotic cells are from 10 to 100 micrometers in diameter. Eucaryotic cells have a much greater volume of cytoplasm and a much lower surface: volume ratio than procaryotic cells. A typical procaryotic cell is about the size of a eucaryotic mitochondrion. Since procaryotes are too small to be seen except with the aid of a microscope, it is usually not appreciated that they are the most abundant form of life on the planet, both in terms of biomass and total numbers of species.

For example, in the sea, procaryotes make up 90 percent of the total combined weight of all organisms. In a single gram of fertile agricultural soil there may be in excess of 10^9 bacterial cells, outnumbering all eucaryotic cells there by 10,000: 1. About 3,000 distinct species of bacteria and archaea are recognized, but this number is probably less than one percent of all the species in nature. These unknown procaryotes, far in excess of undiscovered or unstudied plants, are a tremendous reserve of genetic material and genetic information in nature that awaits exploitation.

Procaryotes are found in all of the habitats where eucaryotes live, but, as well, in many natural environments considered too extreme or inhospitable for eucaryotic cells. Thus, the outer limits of life on Earth (hottest, coldest, driest, etc.) are usually defined by the existence of procaryotes. Where eucaryotes and procaryotes live together, there may be mutualistic associations between the organisms that allow both to survive or flourish. The organelles of eucaryotes (mitochondria and chloroplasts) are thought to be remnants of Bacteria that invaded, or were captured by, primitive eucaryotes in the evolutionary past. Numerous types of eucaryotic cells that exist today are inhabited by endosymbiotic procaryotes.

From a metabolic standpoint, the procaryotes are extraordinarily diverse, and they exhibit several types of metabolism that are rarely or never seen in eucaryotes. For example, the biological processes of nitrogen fixation (conversion of atmospheric nitrogen gas to ammonia) and methanogenesis (production of methane) are metabolically-unique to procaryotes and have an enormous impact on the nitrogen and carbon cycles in nature. Unique mechanisms for energy production and photosynthesis are also seen among the Archaea and Bacteria.

The lives of plants and animals are dependent upon the activities of bacterial cells. Bacteria and archaea enter into various types of symbiotic relationships with plants and animals that usually benefit both organisms, although a few bacteria are agents of disease.

The metabolic activities of procaryotes in soil habitats have an enormous impact on soil fertility that can affect agricultural practices and crop yields. In the global environment, procaryotes are absolutely essential to drive the cycles of elements that make up living systems, i.e., the carbon, oxygen, nitrogen and sulfur cycles. The origins of the plant cell chloroplast and plant-type (oxygenic) photosynthesis are found in procaryotes. Most of the earth's atmospheric oxygen may have been produced by free-living bacterial cells. The bacteria fix nitrogen and a substantial amount of CO_2, as well.

Bacteria or bacterial products (including their genes) can be used to increase crop yield or plant resistance to disease, or to cure or prevent plant disease. Bacterial products include antibiotics to fight infectious disease, as well as components for vaccines used to prevent infectious disease. Because of their simplicity and our relative understanding of their biological processes, the bacteria provide convenient laboratory models for study of the molecular biology, genetics, and physiology of all types of cells, including plant and animal cells.

FUNCTION OF PROKARYOTIC CELLS

Procaryotic cells have three architectural regions: appendages (proteins attached to the cell surface) in the form of flagella and pili; a cell envelope consisting of a capsule, cell wall and plasma membrane; and a cytoplasmic region that contains the cell genome (DNA) and ribosomes and various sorts of inclusions.

SURFACE STRUCTURES

Flagella are filamentous protein structures attached to the cell surface that provide swimming movement for most motile procaryotic cells. The flagellar filament is rotated by a motor apparatus in the plasma membrane allowing the cell to swim in fluid environments. Bacterial flagella are powered by proton motive force (chemiosmotic potential) established on the bacterial membrane, rather than ATP hydrolysis which powers eucaryotic flagella and cilia. Procaryotes are known to exhibit a variety of types of tactic behaviour, i.e., the ability to move (swim) in response to environmental stimuli. For example, during chemotaxis a bacterium can sense the quality and quantity of certain chemicals in their environment and swim towards them (if they are useful nutrients) or away from them (if they are harmful substances).

Fimbriae and Pili are interchangeable terms used to designate short, hair-like structures on the surfaces of procaryotic cells. Fimbriae are shorter and stiffer than flagella, and slightly smaller in diameter. Like flagella, they are composed of protein. A specialized type of pilus the F or sex pilus, mediates the transfer of DNA between mating bacteria, but the function of the smaller, more numerous common pili is quite different. Common pili (almost always called fimbriae) are usually involved in adherence (attachment) of procaryotes to surfaces in nature.

In medical situations, they are major determinants of bacterial virulence because they allow pathogens to attach to (colonize) tissues and to resist attack by phagocytic white blood cells.

THE CELL ENVELOPE

Most procaryotes have a rigid cell wall. The cell wall is an essential structure that protects the delicate cell protoplast from osmotic lysis. The cell wall of Bacteria consists of a polymer of disaccharides cross-linked by short chains of amino acids (peptides). This molecule is a type of peptidoglycan, which is called murein. In the Gram-positive bacteria (those that retain the purple crystal violet dye when subjected to the Gram-staining procedure) the cell wall is a thick layer of murein.

In the Gram-negative bacteria (cells which do not retain the crystal violet dye) the cell wall is relatively thin and is composed of a thin layer of murein surrounded by a membranous structure called the outer membrane. Murein is a substance unique in nature to bacterial cell walls. Also, the outer membrane of Gram-negative bacteria invariably contains a unique component, lipopolysaccharide (LPS or endotoxin), which is toxic to animals. The cell walls of Archaea may be composed of protein, polysaccharides, or peptidgolycan-like molecules, but never do they contain murein. This feature distinguishes the Bacteria from the Archaea.

Although procaryotes lack any intracellular organelles for respiration or photosynthesis, many species possess the physiologic ability to conduct these processes, usually as a function of their plasma membrane. For example, the electron transport system that couples aerobic respiration and ATP synthesis is found in the plasma membrane. The photosynthetic chromophores that harvest light energy for conversion into chemical energy are located in the membrane.

Hence, the plasma membrane is the site of oxidative phosphorylation and photophosphorylation in procaryotes, analogous to the functions of mitochondria and chloroplasts in eucaryotic cells. The procaryotic plasma membrane is also a permeability barrier, and it contains a variety of different transport systems that selectively mediate the passage of substances into and out of the cell.

The membranes of Bacteria are structurally similar to the cell membranes of eucaryotes, except that bacterial membranes consist of saturated or monounsaturated fatty acids (rarely polyunsaturated fatty acids) and do not normally contain sterols. The membranes of Archaea form phospholipid bilayers functionally equivalent to bacterial membranes, but archaeal lipids are saturated, branched, repeating isoprenoid subunits that attach to glycerol via an ether linkage, as opposed to the ester linkage found in glycerides of eucaryotic and bacterial membrane lipids. The structure of archaeal membranes is thought to be an adaptation to their existence in extreme environments.

Most bacteria contain some sort of a polysaccharide layer outside of the cell wall or outer membrane. In a general sense, this layer is called a capsule or glycocalyx. Capsules, slime layers, and glycocalyx are known to mediate attachment of bacterial cells to particular surfaces. Capsules also protect bacteria from engulfment by predatory protozoa or white blood cells (phagocytes) and from attack by antimicrobial agents of plant or animal origin. Capsules in certain soil bacteria protect them from perennial effects of drying or desiccation.

IMPORTANCE OF SURFACE COMPONENTS

All of the various surface components of a procaryotic cell are important in its ecology since they mediate the contact of the cell with its environment. The only "sense" that a procaryote has results from its immediate contact with its environment. It must use its surface components to assess the environment and respond in a way that supports its own existence and survival in that environment. The surface properties of a procaryote are determined by the exact molecular composition of its plasma membrane and cell wall, including LPS, and the function of surface structures such as flagella, fimbriae and capsules. Some important ways that procaryotes use their surface components are:

- As permeability barriers that allow selective passage of nutrients and exclusion of harmful substances;
- As "adhesins" used to attach or adhere to specific surfaces or tissues;
- For protection against engulfment by phagocytic white blood cells or predatory protozoa:
- As enzymes to mediate specific reactions on the cell surface important in the survival of the procaryote;
- As "sensing proteins" that can respond to temperature, osmolarity, salinity, light, oxygen, nutrients, etc. resulting in a signal to the genome of the cell that will cause a biological response to a changing environment.

CYTOPLASMIC CONSTITUENTS

The cytoplasmic constituents of bacteria invariably include the procaryotic chromosome and ribosomes. The chromosome is typically one large circular molecule of DNA, more or less free in the cytoplasm, although intermittently associated with membranes. Procaryotes sometimes possess smaller extrachromosomal pieces of DNA called plasmids. The total DNA content of a cell is referred to as the cell genome. During cell growth and division, the procaryotic chromosome is replicated in the usual semi-conservative fashion before for distribution to progeny cells. However, the eucaryotic processes of meiosis and mitosis are absent in procaryotes. Replication and segregation of procaryotic DNA is coordinated by the plasma membrane.

The distinct granular appearance of procaryotic cytoplasm is due to the presence and distribution of ribosomes. The ribosomes of procaryotes are smaller than cytoplasmic ribosomes of eucaryotes. Procaryotic ribosomes are 70S in size, being composed of 30S and 50S subunits. The 80S ribosomes of eucaryotes are made up of 40S and 60S subunits. Ribosomes are involved in the process of translation (protein synthesis), but some details of their activities differ in eucaryotes, Bacteria and Archaea. Protein synthesis using bacterial 70S ribosomes occurs in eucaryotic mitochondria and chloroplasts, and this is taken as a major line of evidence that these organelles are descended from bacteria.

Often contained in the cytoplasm of procaryotic cells is one or another of some type of inclusion granule. Inclusions are distinct granules that may occupy a substantial part of the cytoplasm. Inclusion granules are usually reserve materials of some sort. For example, carbon and energy reserves may be stored as glycogen (a polymer of glucose) or as polybetahydroxybutyric acid (a type of fat) granules. Polyphosphate inclusions are reserves of PO_4 and possibly energy; elemental sulfur (sulfur globules) are stored by some phototrophic and some lithotrophic procaryotes as reserves of energy or electrons. Some inclusion bodies are actually membranous vesicles or intrusions into the cytoplasm which contain photosynthetic pigments or specialized enzyme complexes.

CLASSIFICATION OF PROCARYOTES

Haeckel (1866) was the first to create a natural Kingdom for the microorganisms, which had been discovered nearly two centuries before by van Leeuwenhoek. He placed all unicellular (microscopic) organisms in a new kingdom, "Protista", separated from plants (Plantae) and animals (Animalia), which were multicellular (macroscopic) organisms. The development of the electron microscope in the 1950's revealed a fundamental dichotomy among Haeckel's "Protista": some cells contained a membrane-enclosed nucleus, and some cells lacked this intracellular structure.

The latter were temporarily shifted to a fourth kingdom, Monera (or Moneres), the procaryotes (also called

Procaryotae). Protista remained as a kingdom of unicellular eucaryotic microorganisms. Whittaker refined the system into five kingdoms in 1967, by identifying the Fungi as a separate multicellular eucaryotic kingdom of organisms, distinguished by their absorptive mode of heterotrophic nutrition.

In the 1980's, Woese began phylogenetic analysis of all forms of cellular life based on comparative sequencing of the small subunit ribosomal RNA (ssrRNA) that is contained in all organisms. A new dichotomy was revealed, this time among the procaryotes: there existed two types of procaryotes, as fundamentally unrelated to one another as they are to eucaryotes. Thus, Woese defined the three cellular Domains of life: Eucarya, Bacteria and Archaea.

Whittaker's Plant, Animal and Fungi kingdoms (all of the multicellular eucaryotes) are at the end of a very small branch of the tree of life, and all other branches lead to microorganisms, either procaryotes (Bacteria and Archaea), or protists (unicellular algae and protozoa), thus establishing clearly that microbial life is the predominant form of life on the planet.

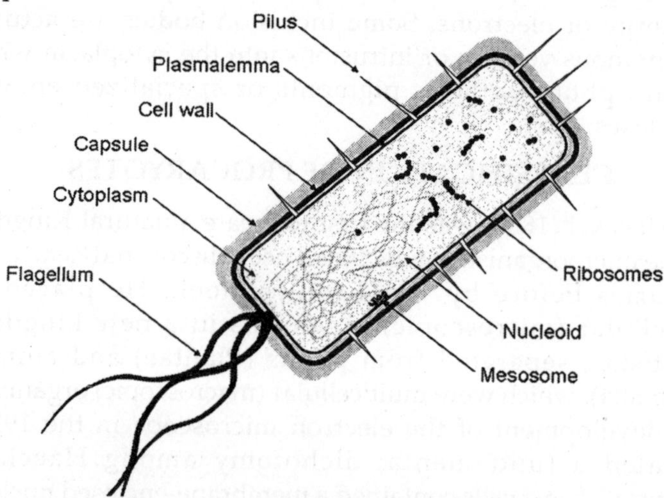

Fig. The Structure of Prokaryotic Cell

Although the definitive difference between Woese's Archaea and Bacteria is based on fundamental differences in the nucleotide base sequence in the 16S ribosomal RNA, there are many biochemical and phenotypic differences between the two groups of procaryotes. The phylogenetic tree indicates that Archaea are more closely related to Eucarya than are Bacteria.

This relatedness seems most evident in the similarities between transcription and translation in the Archaea and the Eucarya. However, it is also evident that the Bacteria have evolved into chloroplasts and mitochondria, so that these eucaryotic organelles derive their lineage from this group of procaryotes. Perhaps the biological success of eucaryotic cells springs from the evolutionary merger of the two procaryotic life forms.

IDENTIFICATION OF BACTERIA

The criteria used for microscopic identification of procaryotes include cell shape and grouping, Gram-stain reaction, and motility. Bacterial cells almost invariably take one of three forms: rod (bacillus), sphere (coccus), or spiral (spirilla and spirochetes). Rods that are curved are called vibrios. Fixed bacterial cells stain either Gram-positive (purple) or Gram-negative (pink); motility is easily determined by observing living specimens. Bacilli may occur singly or form chains of cells; cocci may form chains (streptococci) or grape-like clusters (staphylococci); spiral shape cells are almost always motile; cocci are almost never motile. This nomenclature ignores the actinomycetes, a prominent group of branched bacteria which occur in the soil. But they are easily recognized by their colonies and their microscopic appearance.

Such easily-made microscopic observations, combined with knowing the natural environment of the organism, are important aids to identify the group, if not the exact genus, of a bacterium - providing, of course, that one has an effective key. Such a key is Bergey's Manual of Determinative Bacteriology, the "field guide" to identification of the bacteria. Bergey's Manual describes affiliated groups of Bacteria and Archaea based on a few easily observed microscopic and physiologic characteristics. Further identification requires biochemical tests which will distinguish genera among families and species among genera. Strains within a single species are usually distinguished by genetic or immunological criteria.

REPRODUCTION OF BACTERIA

Most bacteria reproduce by a relatively simple asexual process called binary fission: each cell increases in size and divides into two cells. During this process there is an orderly increase in cellular structures and components, replication and segregation of the bacterial DNA, and formation of a septum or cross wall which divides the cell into two. The process is evidently coordinated by activities associated with the cell membrane. The DNA molecule is believed to be attached to a point on the membrane where it is replicated. The two DNA molecules remain attached at points side-by-side on the membrane while new membrane material is synthesized between the two points.

This draws the DNA molecules in opposite directions while new cell wall and membrane are laid down as a septum between the two chromosomal compartments. When septum formation is complete the cell splits into two progeny cells. The time interval required for a bacterial cell to divide or for a population of cells to double is called the generation time. Generation times for bacterial species growing in nature may be as short as 15 minutes or as long as several days.

GENETIC EXCHANGE IN BACTERIA

Although procaryotes do not undergo sexual reproduction, they are not without the ability to exchange genes and undergo genetic recombination. Bacteria are known to exchange genes in nature by three fundamental processes: conjugation, transduction and transformation. Conjugation involves cell-to-cell contact as DNA crosses a sex pilus from donor to recipient. During transduction, a virus transfers the genes between mating bacteria. In transformation, DNA is acquired directly from the environment, having been released from another cell. Genetic recombination can follow the transfer of DNA from one cell to another leading to the emergence of a new genotype (recombinant).

It is common for DNA to be transferred as plasmids between mating bacteria. Since bacteria usually develop their genes for drug resistance on plasmids (called resistance transfer factors, or RTFs), they are able to spread drug resistance to other strains and species during genetic exchange processes. The genetic engineering of bacterial cells in the research or biotechnology laboratory is often based on the use of plasmids as vectors. The genetic systems of the Archaea are poorly characterized at this point, although the entire genome of *Methanosarcina* has been sequenced which opens up the possibilities for genetic analysis of the group.

EVOLUTION OF BACTERIA

For most procaryotes, mutation is is a major source of variability that allows the species to adapt to new conditions. The mutation rate for most procaryotic genes is in the neighborhood of 10^{-8}.

This means that if a bacterial population doubles from 10^8 cells to 2×10^8 cells, there is likely to be a mutant present for any given gene. Since procaryotes grow to reach population densities far in excess of 10^9 cells, such a mutant could develop from a single generation during 15 minutes of growth. The evolution of procaryotes, driven by such Darwinian principles of evolution (mutation and selection) is called vertical evolution.

However, as a result of the processes of genetic exchange described above, the bacteria and archaea can also undergo a process of horizontal evolution. In this case, genes are transferred laterally from one organism to another, even between members of different Kingdoms, which allows the recipient to experiment with a new genetic trait. Horizontal evolution is becoming realized to be a significant force in driving cellular evolution.

The combined effects of fast growth rates, high concentrations of cells, genetic processes of mutation and selection, and the ability to exchange genes, account for the extraordinary rates of adaptation and evolution that can be observed in the procaryotes.

ECOLOGY OF BACTERIA

Bacteria and Archaea are present in all environments that support life. They may be free-living, or living in associations with "higher forms" of life (plants and animals), and they are found in environments that support no other form of life. Procaryotes have the usual nutritional requirements for growth of cells, but many of the ways that they utilize and transform their nutrients are unique. This bears directly on their habitat and their ecology.

In terms of carbon utilization a cell may be heterotrophic or autotrophic. Heterotrophs obtain their carbon and energy for growth from organic compounds in nature. Autotrophs use $C0_2$ as a sole source of carbon for growth and obtain their energy from light (e.g. photoautotrophs) or from the oxidation of inorganic compounds (e.g. lithoautotrophs).

Most heterotrophic bacteria are saprophytes, meaning that they obtain their nourishment from dead organic matter. In the soil, saprophytic bacteria and fungi are responsible for biodegradation of organic material. Ultimately, organic molecules, no matter how complex, can be degraded to CO_2 (plus H_2 and H_2O).Probably no naturally-occurring organic substance cannot be degraded by the combined activities of the bacteria and fungi. Hence, most organic matter in nature is converted by heterotrophs to CO_2, only to be converted back into organic material by autotrophs that die and nourish heterotrophs to complete the carbon cycle.

Lithotrophic procaryotes have a type of energy-producing metabolism which is unique. Lithotrophs (also called lithoautotrophs or chemoautotrophs) use inorganic compounds as sources of energy, i.e., they oxidize compounds such as H_2 or H_2S or NH_3 to obtain electrons to feed in to an electron transport system and to produce ATP.

Lithotrophs are found in soil and aquatic environments wherever their energy source is present. Most lithotrophs are autotrophs so they can grow in the absence of any organic material. Lithotrophic species are found among the Bacteria and the Archaea. Sulfur-oxidizing lithotrophs convert H_2S to S^o and S^o to SO_4. Nitrifying bacteria convert NH_3 to NO_2 and

NO_2 to NO_3; methanogenic archaea strip electrons off of H_2 as a source of energy and add them to CO_2 to form CH_4 (methane). Lithotrophs have an obvious impact on the sulfur, nitrogen and carbon cycles in the biosphere.

Photosynthetic bacteria convert light energy into chemical energy for growth. Most phototrophic bacteria are autotrophs so their role in the carbon cycle is analogous to that of plants. The planktonic cyanobacteria are the "grass of the sea" and their form of oxygenic photosynthesis generates a substantial amount of O_2 in the biosphere. However, among the photosynthetic bacteria are types of photosynthetic metabolism not seen in eucaryotes, including photohetero-trophy (using light as an energy source while assimilating organic compounds as a source of carbon), anoxygenic photosynthesis, and unique mechanisms of CO_2 fixation (autotrophy).

Photosynthesis has not been found to occur among the Archaea, but one archaeal species employs a light-driven non photosynthetic means of energy generation based on the use of a chromophore called bacteriorhodopsin.

Adaptations to Environmental Conditions

Most procaryotes, whether they have been cultured and studied in the laboratory, or observed growing in their natural habitats, seem to be highly adapted to their specific environment by means of their macromolecular structure and/or their physiologic (metabolic) capabilities. The nutritional quality of the environment determines whether a particular organism will be present, but so do various physical parameters such as the availability of light and O_2, as well as the pH, temperature and salinity of the environment. As examples, the range of procaryotic responses to oxygen and temperature are discussed below.

Procaryotes vary widely in their response to O_2 (molecular oxygen). Organisms that require O_2 for growth are called obligate aerobes; those which are inhibited or killed by O_2, and which grow only in its absence, are called obligate anaerobes; organisms which grow either in the presence or absence of O_2 are called facultative anaerobes. Whether or not a particular organism can exist in the presence of O_2 depends upon the distribution of certain enzymes such as superoxide dismutase and catalase that are required to detoxify lethal oxygen radicals that are always generated by living systems in the presence of O_2.

Procaryotes also vary widely in their response to temperature. Those that live at very cold temperatures (0 degrees or lower) are called psychrophiles; those which flourish at room temperature (25 degrees) or at the temperature of warm-blooded animals (37 degrees) are called mesophiles; those that live at high temperatures (greater than 45 degrees) are thermophiles. The only limit that seems to be placed on growth of certain procaryotes in nature relative to temperature is whether liquid water exists.

Hence, growing procaryotic cells can be found in supercooled environments (ice does not form) as low as -20 degrees and superheated environments (steam does not form) as high as 120 degrees. Archaea have been detected around thermal vents on the ocean floor where the temperature is as high as 320 degrees!

BACTERIAL SYMBIOSIS

The biomass of procaryotic cells in the biosphere, their metabolic diversity, and their persistence in all habitats that support life, ensures that these microbe will play a crucial role in the cycles of elements and the functioning of the world ecosystem. However, the procaryotes affect the world ecology in another significant way through their inevitable interactions with insects, plants and animals. Some bacteria are required to associate with insects, animals or plants for the latter to survive.

For example, the sex of offspring of certain insects is determined by endosymbiotic bacteria. Ruminant animals (cows, sheep, etc.), whose diet is mainly cellulose (plant material), must have cellulose-digesting bacteria in their intestine to convert the cellulose to a form of carbon that the animal can assimilate. Leguminous plants grow poorly in nitrogen-deprived soils unless they are colonized by nitrogen-fixing bacteria which can supply them with a biologically-useful form of nitrogen.

BACTERIAL PATHOGENICITY

Such bacteria that cause disease in plants or animals are pathogens. Some bacteria are parasites of plants or animals, meaning that they grow at the expense of their eucaryotic host and may damage, harm, or even kill it in the process. Human diseases caused by bacterial pathogens include tuberculosis, whooping cough, diphtheria, tetanus, gonorrhea, syphilis, pneumonia, cholera and typhoid fever, to name a few. The bacteria that cause these diseases have special structural or biochemical properties that determine their virulence or pathogenicity. These include:

- Ability to colonize and invade their host;
- Ability to resist or withstand the antibacterial defenses of the host; ability to produce various toxic substances that damage the host. Plant diseases, likewise, may be caused by bacterial pathogens. More than 200 species of bacteria are associated with plant diseases, but a very small handful of genera are involved.

BACTERIAL APPLICATIONS

Procaryotes, especially bacteria, are used industrially in the manufacture of foods, antibiotics, drugs, vaccines, insecticides, enzymes, hormones and other useful biological products. In fact, through genetic engineering of bacteria, these unicellular organisms can be coaxed to produce just about anything that there is a gene to encode for. The genetic systems of bacteria are the foundation of the biotechnology industry. In the foods industry, lactic acid bacteria such as *Lactobacillus* and *Streptococcus* are used the manufacture of dairy products such as yogurt, cheese, buttermilk, sour cream, and butter.

Lactic acid fermentations are also used in pickling processes. Bacterial fermentations can be used to produce lactic acid, acetic acid, ethanol or acetone. In many parts of the world, various human cultures ferment indigenous plant material using *Zymomonas* bacteria to produce the regional alcoholic beverage. For example, in Mexico, a Maguey cactus (*Agave*) is fermented to "cactus beer" or pulque. Pulque can be ingested as is, or distilled into tequila.

In the pharmaceutical industry, bacteria are used to produce antibiotics, vaccines, and medically-useful enzymes. Most antibiotics are made by bacteria that live in soil. Actinomycetes such as *Streptomyces* produce tetracyclines, erythromycin, streptomycin, rifamycin and ivermectin. *Bacillus* species produce bacitracin and polymyxin. Bacterial products are used in the manufacture of vaccines for immunization against infectious disease.

Vaccines against diphtheria, whooping cough, tetanus, typhoid fever and cholera are made from components of the bacteria that cause the respective diseases. It is significant to note here that the use of antibiotics against infectious disease and the widespread practice of vaccination (immunization) against infectious disease are two twentieth-century developments that have drastically increased the quality of life and the average life expectancy of individuals in developed countries.

Chapter 5

World of Fungi

Study of Fungi is calle Mycology. The genetic and biochemical properties of Fungi, their taxonomy, and their use to humans as a source for tinder, medicinals (e.g., penicillin), food (e.g., beer, wine, cheese, edible mushrooms) and entheogens, as well as their dangers, such as poisoning or infection. From mycology arose the field of phytopathology, the study of plant diseases, and the two disciplines remain closely related because the vast majority of plant pathogens are fungi. A biologist who studies mycology is called a mycologist.

Historically, mycology was a branch of botany (fungi are evolutionarily more closely related to animals than to plants but this was not recognized until a few decades ago). Pioneer mycologists included Elias Magnus Fries, Christian Hendrik Persoon, Anton de Bary and Lewis David von Schweinitz. Today, the most comprehensively studied and understood fungi are the yeasts and eukaryotic model organisms Saccharomyces cerevisiae and Schizosaccharomyces pombe.

Many fungi produce toxins, antibiotics, and other secondary metabolites. For example, the cosmopolitan (worldwide) genus Fusarium and their toxins associated with fatal outbreaks of alimentary toxic aleukia in humans were extensively studied by Abraham Joffe. Fungi are fundamental for life on earth in their roles as symbionts, e.g. in the form of mycorrhizae, insect symbionts and lichens as well as their potency in breaking down complex organic biomolecules such as lignin, the more durable component of wood, as well as xenobiotics, a critical step in the global carbon cycle.

Fungi and other organisms traditionally recognized as fungi, such as oomycetes and myxomycetes (slime molds), often are economically and socially important as some cause diseases of animals (such as histoplasmosis) as well as plants (such as Dutch elm disease and Rice blast).

Field meetings to find interesting species of fungi are known as 'forays', after the first such meeting organized by the Woolhope Naturalists' Field Club in 1868 and entitled "a foray among the fungi." Some fungi can cause disease in humans or other organisms. The study of pathogenic fungi is referred to as medical mycology. Fungi are eukaryotic organisms that do not contain chlorophyll, but have cell walls, filamentous structures, and produce spores. These organisms grow as saprophytes and decompose dead organic matter. There are between 100,000 to 200,000 species depending on how they are classified. About 300 species are presently known to be pathogenic for man.

There are four types of mycotic diseases:

- *Hypersensitivity:* An allergic reaction to molds and spores.
- *Mycotoxicoses:* Poisoning of man and animals by feeds and food products contaminated by fungi which produce toxins from the grain substrate.
- *Mycetismus:* The ingestion of preformed toxin (mushroom poisoning).
- *Infection:* In this section, we shall be concerned only with the last type.

MORPHOLOGY

A mass of hyphae is called mycelia. Yeasts are unicellular organisms and mycelia are multicellular filamentous structures, constituted by tubular cells with cell walls. The yeasts reproduce by budding. The mycelial forms branch and the pattern of branching is an aid to the morphological identification. If the mycelia do not have *septa,* they are called coenocytic (nonseptate). The terms "hypha" and "mycelium" are frequently used interchangeably. Some fungi occur in both the yeast and mycelial forms. These are called dimorphic fungi.

Types of Dimorphic fungi

There are two forms of Dimorphic Fungi:

- Mycelium
- Yeast

Mycelium (Saprophytic Form): The form observed in nature or when cultured at 25 degrees C. Conversion to the yeast form appears to be essential for pathogenicity. In the dimorphic fungi. Fungi are identified by several morphological or biochemical characteristics, including the appearance of their fruiting bodies. The asexual spores may be large (macroconidia, chlamydospores) or small (microconidia, blastospores, arthroconidia).

Yeast (parasitic or pathogenic form): This is the form usually seen in tissue, in exudates, or if cultured in an incubator at 37 degrees C.

Types of Mycotic Diseases:

- *Hypersensitivity:* An allergic reaction to molds and spores.
- *Mycotoxicoses:* Poisoning of man and animals by feeds and food products contaminated by fungi which produce toxins from the grain substrate.
- *Mycetismus:* The ingestion of toxin (mushroom poisoning).
- *Infection:* We shall be concerned only with the last type: pathogenic fungi that cause infections.

Most common pathogenic fungi do not produce toxins but they do show physiologic modifications during a parasitic infection (e.g., increased metabolic rate, modified metabolic pathways and modified cell wall structure). The mechanisms that cause these modifications as well as their significance as a pathogenic mechanism are just being described. Most pathogenic fungi are also thermotolerant, and can resist the effects of the active oxygen radicals released during the respiratory burst of phagocytes.

Thus, fungi are able to withstand many host defenses. Fungi are ubiquitous in nature and most people are exposed to them. The establishment of a mycotic infection usually depends on the size of the inoculum and on the resistance of the host. The severity of the infection seems to depend mostly on the immunologic status of the host. Thus, the demonstration of fungi, for example, in blood drawn from an intravenous catheter can correspond to colonization of the catheter, to transient fungemia (i.e., dissemination of fungi through the blood strem), or to a true infection.

The physician must decide which is the clinical status of the patient based on clinical parameters, general status of the patient, laboratory results, etc. The decision is not trivial, since treatment of systemic fungal infections requires the aggressive use of drugs with considerable toxicity. Most mycotic agents are soil saprophytes and mycotic diseases are generally not communicable from person-to-person (occasional exceptions: Candida and some dermatophytes).

Outbreaks of disease may occur, but these are due to a common environmental exposure, not communicability. Most of the fungi which cause systemic infections have a peculiar, characteristic ecologic niche in nature. This habitat is specific for several fungi which will be discussed later. In this environment, the normally saprophytic organisms proliferate and develop. This habitat is also the source of fungal elements and/or spores, where man and animals, incidental hosts, are exposed to the infectious particles.

It is important to be aware of these associations to diagnose mycotic diseases. The physician must be able to elicit a complete history from the patient including occupation, avocation and travel history. This information is frequently required to raise, or confirm, your differential diagnosis. The incidence of mycotic infections is currently increasing dramatically, due to an increased population of susceptibles. Examples are patients with AIDS, patients on immunosuppressive therapy, and the use of more invasive diagnostic and surgical procedures (prosthetic implants). Fungal diseases are non-contagious and non-reportable diseases in the national public health statistics. However, in South Carolina most of the important mycotic (fungal) diseases were notifiable to the public health authorities until 1994.

CLASSIFICATION OF FUNGI

Fungi are eukaryotic organisms that do not contain chlorophyll, but have cell walls, filamentous structures, and produce spores. These organisms grow as saprophytes and decompose dead organic matter. There are between 100,000 to 200,000 species depending on how they are classified. About 300 species are presently known to be pathogenic for man.

There are five kingdoms of living things. The fungi are in the Kingdom Fungi. This common characteristic is responsible for the therapeutic dilemma in anti-mycotic therapy. The taxonomy of the Kingdom Fungi is evolving and is controversial. Formerly based on gross and light microscopic morphology, studies of ultra structure, biochemistry and molecular biology provide new evidence on which to base taxonomic positions. Medically important fungi are in four phyla:

- *Ascomycota:* Sexual reproduction in a sack called an ascus with the production of ascopspores.
- *Basidiomycota:* Sexual reproduction in a sack called a basidium with the production of basidiospores.
- *Zygomycota:* Sexual reproduction by gametes and asexual reproduction with the formation of zygospores.
- *Mitosporic Fungi (Fungi Imperfecti):* No recognizable form of sexual reproduction. Includes most pathogenic fungi.

DIAGNOSIS

Skin testing (dermal hypersensitivity) used to be popular as a diagnostic tool, but this use is now discouraged because the skin test may interfere with serological studies, by causing false positive results. It may still be used to evaluate the patient's immunity, as well as a population exposure index in epidemiological studies.

Skin scrapings suspected to contain dermatophytes or pus from a lesion can be mounted in KOH on a slide and examined directly under the microscope.

Direct fluorescent microscopy may be used for

identification, even on non-viable cultures or on fixed tissue sections. The reagents for this test are difficult to obtain.

Serology may be helpful when it is applied to a specific fungal disease; there are no screening antigens for 'fungi' in general. Because fungi are poor antigens, the efficacy of serology varies with different fungal infections.

The serologic tests will be discussed under each mycosis. The most common serological tests for fungi are based on latex agglutination, double immunodiffusion, complement fixation and enzyme immunoassays.

While latex agglutination may favour the detection of IgM antibodies, double immunodiffusion and complement fixation usually detect IgG antibodies. Some EIA tests are being developed to detect both IgG and IgM antibodies. There are some tests which can detect specific fungal antigens, but they are just coming into general use.

Biopsy and histopathology. A biopsy may be very useful for the identification and as a source of the of tissue-invading fungi. Usually the Gomori methenamine silver (GMS) stain is used to reveal the organisms which stain black against a green background. The H&E stain does not always tint the organism, but it will stain the inflammatory cells.

Culture. A definitive diagnosis requires a culture and identification. Pathogenic fungi are usually grown on Sabouraud dextrose agar. It has a slightly acidic pH (~5.6); cyclohexamide, penicillin, streptomycin or other inhibitory antibiotics are often added to prevent bacterial contamination and overgrowth. Two cultures are inoculated and incubated separately at 25 degrees C and 37 degrees C to reveal dimorphism. The cultures are examined macroscopically and microscopically. They are not considered negative for growth until after 4 weeks of incubation.

TREATMENT

The pathogens are difficult to eradicate by the animal host defence mechanisms. Because mammals and fungi are both eukaryotic, the cellular milieu is biochemically similar in both. The cell membranes of all eukaryotic cells contain sterols;

ergosterol in the fungal cell membrane and cholesterol in the mammalian cell membrane.

Thus, most substances which may impair the invading fungus will usually have serious side effects on the host. Although one of the first chemotherapeutic agents (oral iodides) was an anti-mycotic used in 1903, the further development of such agents has been left far behind the development of anti-bacterial agents.

CHEMICALCCONTROL OF FUNGUS

Griseofulvin

Griseofulvin is a very slow-acting drug which is used for severe skin and nail infections. Its effect depends on its accumulation in the stratum corneum where it is incorporated into the tissue and forms a barrier which stops further fungal penetration and growth. It is administered orally. The exact mechanism of action is unknown.

5-fluorocytosine

5-fluorocytosine (Flucytosine or 5-FC) inhibits RNA synthesis and has found its main application in cryptococcosis (to be discussed later). It is administered orally.

Amphotericin B

A polyene antimycotic. It is usually the drug of choice for most systemic fungal infections. It has a greater affinity for ergosterol in the cell membranes of fungi than for the cholesterol in the host's cells; once bound to ergosterol, it causes disruption of the cell membrane and death of the fungal cell. Amphotericin B is usually administered intravenously (patient usually needs to be hospitalized), often for 2-3 months. The drug is rather toxic; thrombo-phlebitis, nephrotoxicity, fever, chills and anemia frequently occur during administration.

Azoles

The azoles (imidazoles and triazoles), including

ketoconazole, fluconazole, and itraconozole, are being used for muco-cutaneous candidiasis, dermatophytosis, and for some systemic fungal infections. Fluconazole is presently essential for the maintenance of AIDS patients with cryptococcosis. The general mechanism of action of the azoles is the inhibition of ergosterol synthesis. Oral administration and reduced toxicity are distinct advantages.

MEDICAL CLASSIFICATION OF FUNGAL DISEASES

In Clinical Terms Fungal diseases may be discussed in a variety of ways. The most practical method for medical students is the clinical taxonomy which divides the fungi into:

- Opportunistic mycoses
- Subcutaneous mycoses
- Systemic mycoses
- Superficial mycoses

The Opportunistic mycoses are infections due to fungi with low inherent virulence. The etiologic agents are organisms which are common in all environments. The Superficial mycoses (or cutaneous mycoses) are fungal diseases that are confined to the outer layers of the skin, nail, or hair, (keratinized layers) rarely invading the deeper tissue or viscera. The fungi involved are called dermatophytes.

The Subcutaneous mycoses are confined to the subcutaneous tissue and only rarely spread systemically. They usually form deep, ulcerated skin lesions or fungating masses, most commonly involving the lower extremities. The causative organisms are soil saprophytes which are introduced through trauma to the feet or legs. The Systemic mycoses may involve deep viscera and become widely disseminated. Each fungus type has its own predilection for various organs which will be described as we discuss the individual diseases.

GENERATION OF ACTINOMYCETES

There are three generation of actinomycetes:

- Actinomyces,
- Nocardia,
- Streptomyces.

These organisms have been shown to be higher bacteria, but they were thought to be fungi for many years because they have filamentous forms, 0.5 to 0.8 microns in diameter, which appear to branch. Some species form aerial mycelia in culture. The clinical manifestations of infection are similar to those of a systemic fungal infection. It is now clear that they are not fungi but are closely related to the mycobacteria.

Facts about these generation:

- Actinomyces are anaerobic, while Nocardia and Streptomyces are aerobic.
- Nocardia stain partially acid-fast, Actinomyces and Streptomyces are not acid-fast.
- Actinomyces produce granules. Most actinomycetes in tissue do not stain with the H & E stain commonly used for general histopathology. All genera may produce granules; Actinomyces almost always produce granules.

Actinomycosis

Actinomycosis is a chronic suppurative and granulomatous disease of the cervico-facial, thoracic or abdominal areas. The most common cause of actinomycosis is the organism *Actinomyces israelii* which infects both man and animals. In cattle, the disease is called "lumpy jaw" because of the huge abscess formed in the angle of the jaw. In man, *A. israelii* is an endogenous organism that can be isolated from the mouths of healthy people.

Frequently, the infected patient has a tooth abscess or a tooth extraction and the endogenous organism becomes established in the traumatized tissue and causes a suppurative infection. These abscesses are not confined to the jaw and may also be found in the thoracic area and abdomen. The patient usually presents with a pus-draining lesion, so the pus will be the clinical material you send to the laboratory. This diagnosis can be made on the hospital floor. If you rotate the vial of pus, the yellow sulfur granules, characteristic of this organism, can be seen with the naked eye. You can also see these granules by running sterile water over the gauze used to cover the lesion. The water washes away the purulent material leaving the golden granules on the gauze. This organism, which occurs worldwide, can be seen histologically as "sulfur granules" surrounded by polymorphonuclear cells (PMN) forming the purulent tissue reaction.

The organism is a gram positive rod that frequently branches. The laboratory must specifically be instructed to culture for this anaerobic organism. These lesions must be surgically drained prior to antibiotic therapy and the drug of choice is large doses of penicillin.

STREPTOMYCOSIS

The streptomyces species usually cause the disease entity known as mycetoma (fungus tumor). These infections are usually subcutaneous, but they can penetrate deeper and invade the bone. Some species produce a protease which inhibits macrophages. Material sent to the lab is pus or skin biopsy. The streptomycetes are aerobic like Nocardia, and can grow on both bacterial and fungal (Sabouraud) media. They produce a chalky aerial mycelium with much branching. It is important to let the lab know the organism you suspect because most bacterial pathogens will grow out overnight, but the actinomycetes take longer to be visible on the culture plates (48-72 hours).

The various species of streptomyces produce granules of different size texture and colour. These granules along with colonial growth and biochemical tests allow the bacteriologist or mycologist to identify each species. The organisms are found world-wide. There are no serological tests, and the drugs of choice are the combination of sulfamethoxazole/ trimethoprim or amphotericin B. In the tropics this disease may go undiagnosed or untreated for so long that surgical amputation may be the only effective treatment.

NOCARDIOSIS

Nocardiosis primarily presents as a pulmonary disease or brain abscess in the U.S. In Latin America, it is more frequently seen as the cause of a subcutaneous infection, with or without draining abscesses. It can even present as a lesion in the chest wall that drains onto the surface of the body similar to actinomycosis. Brain abscesses are frequent secondary lesions.

The most common species of Nocardia which cause disease in human beings are *N. brasiliensis* and *N. asteroides*. These are soil organisms which can also be found endogenously in the sputum of apparently healthy people. *N. asteroides* is usually the etiologic agent of pulmonary nocardiosis (9-11) while *N. brasiliensis* (8B) is frequently the cause of sub-cutaneous lesions. The material sent to the lab, depending on the presentation of the disease, is sputum, pus, or biopsy material.

These organisms rarely form granules. The Nocardia are aerobic, gram-positive rods and stain partially acid-fast (i.e., the acid-fast staining is not uniform). There are no serological tests, and the drug of choice is Bactrim (Trimethoprim plus sulfamethoxazole). The nocardia grow readily on most bacteriologic and TB media.

Chapter 6

Study of Virus

The study of viruses is called Virology and their structure, classification and evolution, their ways to infect and exploit cells for virus reproduction, the diseases they cause, the techniques to isolate and culture them, and their use in research and therapy.

VIRUS STRUCTURE AND CLASSIFICATION

A major branch of virology is virus classification. Viruses can be classified according to the host cell they infect: animal viruses, plant viruses, fungal viruses, and bacteriophages (viruses infecting bacteria, which include the most complex viruses). Another classification uses the geometrical shape of their capsid (often a helix or an icosahedron) or the virus's structure (e.g. presence or absence of a lipid envelope). Viruses range in size from about 30 nm to about 450 nm, which means that most of them cannot be seen with light microscopes. The shape and structure of viruses can be studied with electron microscopy, with NMR spectroscopy, and most importantly with X-ray crystallography.

The most useful and most widely used classification system distinguishes viruses according to the type of nucleic acid they use as genetic material and the viral replication method they employ to coax host cells into producing more viruses:

- DNA viruses (divided into double-stranded DNA viruses and the much less common single-stranded DNA viruses),
- RNA viruses (divided into positive-sense single-

stranded RNA viruses, negative-sense single-stranded RNA viruses and the much less common double-stranded RNA viruses),

- Reverse transcribing viruses (double-stranded reverse-transcribing DNA viruses and single-stranded reverse-transcribing RNA viruses including retroviruses).

In addition virologists also study *subviral particles*, infectious entities even smaller than viruses: viroids (naked circular RNA molecules infecting plants), satellites (nucleic acid molecules with or without a capsid that require a helper virus for infection and reproduction), and prions (proteins that can exist in a pathological conformation that induces other prion molecules to assume that same conformation).

The latest report by the International Committee on Taxonomy of Viruses (2005) lists 5450 viruses, organized in over 2,000 species, 287 genera, 73 families and 3 orders.

The taxa in virology are not necessarily monophyletic. In fact, the evolutionary relationships of the various virus groups remain unclear, and three hypotheses regarding their origin exist:

- Viruses arose from non-living matter, separately from and in parallel to other life forms, possibly in the form of self-reproducing RNA ribozymes similar to viroids.
- Viruses arose from earlier, more competent cellular life forms that became parasites to host cells and subsequently lost most of their functionality; examples of such tiny parasitic prokaryotes are Mycoplasma and Nanoarchaea.
- Viruses arose as parts of the genome of cells, most likely transposons or plasmids, that acquired the ability to "break free" from the host cell and infect other cells.

It is of course possible that different alternatives apply to different virus groups.

Of particular interest here is mimivirus, a giant virus that infects amoebae and carries much of the molecular machinery traditionally associated with bacteria. Is it a simplified version of a parasitic prokaryote, or did it originate as a simpler virus that acquired genes from its host? The evolution of viruses, which often occurs in concert with the evolution of their hosts, is studied in the field of viral evolution.

While viruses reproduce and evolve, they don't engage in metabolism and depend on a host cell for reproduction. The often-debated question of whether they are alive or not is a matter of definition that does not affect the biological reality of viruses.

VIRAL DISEASES AND HOST DEFENSES

One main motivation for the study of viruses is the fact that they cause many important infectious diseases, among them the common cold, influenza, rabies, measles, many forms of diarrhea, hepatitis, yellow fever, polio, smallpox and AIDS. Some viruses, known as oncoviruses, contribute to certain forms of cancer; the best studied example is the association between Human papillomavirus and cervical cancer. Some subviral particles also cause disease: Kuru and Creutzfeldt-Jakob disease are caused by prions, and hepatitis D is due to a satellite virus.

The study of the manner in which viruses cause disease is viral pathogenesis. The degree to which a virus causes disease is its virulence.

When the immune system of a vertebrate encounters a virus, it produces specific antibodies which bind to the virus and mark it for destruction. The presence of these antibodies is often used to determine whether a person has been exposed to a given virus in the past, with tests such as ELISA. Vaccinations protect against viral diseases, in part, by eliciting the production of antibodies. Specifically constructed monoclonal antibodies can also be used to detect the presence of viruses, with a technique called fluorescence microscopy.

A second defense of vertebrates against viruses, cell-mediated immunity, involves immune cells known as T cells: the body's cells constantly display short fragments of their proteins on the cell's surface, and if a T cell recognizes a suspicious viral fragment there, the host cell is destroyed and the virus-specific T-cells proliferate. This mechanism is jump-started by certain vaccinations.

RNA interference, an important cellular mechanism found in plants, animals and many other eukaryotes, most likely evolved as a defense against viruses. An elaborate machinery of interacting enzymes detects double-stranded RNA molecules (which occur as part of the life cycle of many viruses) and then proceeds to destroy all single-stranded versions of those detected RNA molecules.

Every lethal viral disease presents a paradox: killing its host is obviously of no benefit to the virus, so how and why did it evolve to do so? Today it is believed that most viruses are relatively benign in their natural hosts; the lethal viral diseases are explained as resulting from an "accidental" jump of the virus from a species in which it is benign to a new one that is not accustomed to it. For example, serious influenza viruses probably have pigs or birds as their natural host, and HIV is thought to derive from the benign monkey virus SIV.

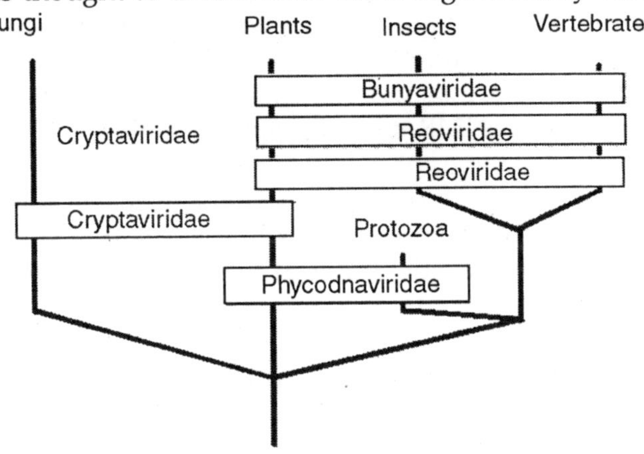

Fig. Virus Classification

While it has been possible to prevent (certain) viral diseases by vaccination for a long time, the development of antiviral drugs to *treat* viral diseases is a comparatively recent development. The first such drug was interferon, a substance that is naturally produced by certain immune cells when an infection is detected and stimulates other parts of the immune system.

Virology, often considered a part of microbiology or of pathology, is the study of biological viruses and virus like agents: their structure and classification, their ways to infect and exploit cells for virus reproduction, the diseases they cause, the techniques to isolate and culture them, and their potential uses in research and therapy. A researcher in virology is a virologist.

STRUCTURE AND CLASSIFICATION OF VIRUS

A major branch of virology is virus classification. Viruses can be classified according to the host cell they infect: animal viruses, plant viruses, fungal viruses, and bacteriophages (viruses infecting bacteria, which include the most complex viruses). Another classification uses the geometrical shape of their capsid (often a helix or an icosahedron) or the virus's structure (e.g. presence or absence of a lipid envelope). Viruses range in size from about 30 nm to about 450 nm, which means that most of them cannot be seen with light microscopes. The shape and structure of viruses can be studied with electron microscopy, with NMR spectroscopy, and most importantly with X-ray crystallography.

The most useful and most widely used classification system distinguishes viruses according to the type of nucleic acid they use as genetic material and the viral replication method they employ to coax host cells into producing more viruses:

- DNA viruses (divided into double-stranded DNA viruses and the much less common single-stranded DNA viruses),
- RNA viruses (divided into positive-sense single-stranded RNA viruses, negative-sense single-stranded RNA viruses and the much less common double-stranded RNA viruses),
- Reverse transcribing viruses (double-stranded reverse-transcribing DNA viruses and single-

stranded reverse-transcribing RNA viruses including retroviruses).

In addition virologists also study *subviral particles*, infectious entities even smaller than viruses: viroids (naked circular RNA molecules infecting plants), satellites (nucleic acid molecules with or without a capsid that require a helper virus for infection and reproduction), and prions (proteins that can exist in a conformation which induces other protein molecules to assume that same conformation).

The latest report by the International Committee on Taxonomy of Viruses (2005) lists 5450 viruses, organized in over 2,000 species, 287 genera, 73 families and 3 orders.

The taxa in virology are not necessarily monophyletic. In fact, the evolutionary relationships of the various virus groups remain unclear, and three hypotheses regarding their origin exist:

- Viruses arose from non-living matter, separately from and in parallel to other life forms, possibly in the form of self-reproducing RNA ribozymes similar to viroids.
- Viruses arose from earlier, more competent cellular life forms that became parasites to host cells and subsequently lost most of their functionality; examples of such tiny parasitic prokaryotes are Mycoplasma and Nanoarchaea.
- Viruses arose as parts of the genome of cells, most likely transposons or plasmids, that acquired the ability to "break free" from the host cell and infect other cells.

It is of course possible that different alternatives apply to different virus groups. Of particular interest here is mimivirus, a giant virus that infects amoebae and carries much of the molecular machinery traditionally associated with bacteria. Is it a simplified version of a parasitic prokaryote, or did it originate as a simpler virus that acquired genes from its host?

While viruses reproduce and evolve, they don't engage in metabolism and depend on a host cell for reproduction. The often-debated question of whether they are alive or not is a matter of definition that does not affect the biological reality of viruses.

VIRAL DISEASES AND HOST DEFENSES

One main motivation for the study of viruses is the fact that they cause many important infectious diseases, among them the common cold, influenza, rabies, measles, many forms of diarrhea, hepatitis, yellow fever, polio, smallpox and AIDS. Some viruses, known as oncoviruses, contribute to certain forms of cancer; the best studied example is the association between Human papillomavirus and cervical cancer. Some subviral particles also cause disease: Kuru and Creutzfeldt-Jakob disease are caused by prions, and hepatitis D is due to a satellite virus.

The study of the manner in which viruses cause disease is viral pathogenesis. The degree to which a virus causes disease is its virulence.

When the immune system of a vertebrate encounters a virus, it produces specific antibodies which bind to the virus and mark it for destruction. The presence of these antibodies is often used to determine whether a person has been exposed to a given virus in the past, with tests such as ELISA. Vaccinations protect against viral diseases, in part, by eliciting the production of antibodies. Specifically constructed monoclonal antibodies can also be used to detect the presence of viruses, with a technique called fluorescence microscopy.

A second defence of vertebrates against viruses, cell-mediated immunity, involves immune cells known as T cells: the body's cells constantly display short fragments of their proteins on the cell's surface, and if a T cell recognizes a supicious viral fragment there, the host cell is destroyed and the virus-specific T-cells proliferate. This mechanism is jump-started by certain vaccinations.

RNA interference, an important cellular mechanism found in plants, animals and many other eukaryotes, most likely evolved as a defence against viruses. An elaborate machinery of interacting enzymes detects double-stranded RNA molecules (which occur as part of the life cycle of many viruses) and then proceeds to destroy all single-stranded versions of those detected RNA molecules.

Every lethal viral disease presents a paradox: killing its host is obviously of no benefit to the virus, so how and why did it evolve? Today it is believed that most viruses are relatively benign in their natural host; the lethal viral diseases are explained as resulting from an "accidental" jump of the virus from a species in which it is benign to a new one that is not accustomed to it. For example, serious influenza viruses probably have pigs or birds as their natural host, and HIV is thought to derive from the benign monkey virus SIV.

While it has been possible to prevent (certain) viral diseases by vaccination for a long time, the development of antiviral drugs to *treat* viral diseases is a comparatively recent development. The first such drug was interferon, a substance that is naturally produced by certain immune cells when an infection is detected, thus stimulating other parts of the immune system.

VIRAL THERAPY

Bacteriophages, the viruses which infect bacteria, can be relatively easily grown as viral plaques on bacterial cultures. Bacteriophages occasionally move genetic material from one bacterial cell to another in a process known as transduction, and this horizontal gene transfer is one reason why they served as a major research tool in the early development of molecular biology. The genetic code, the function of ribozymes, the first recombinant DNA and early genetic libraries were all arrived at using bacteriophages. Certain genetic elements derived from viruses, such as highly effective promoters, are commonly used in molecular biology research today.

Growing animal viruses outside of the living host animal is more difficult. Classically, fertilized chicken eggs have often been used, but cell cultures are increasingly employed for this purpose today.

Since viruses that infect eukaryotes need to transport their genetic material into the host cell's nucleus, they are attractive tools for introducing new genes into the host (known as transformation or transfection), and this approach of using viruses as gene vectors is being pursued in the gene therapy of genetic diseases. An obvious problem to be overcome in viral gene therapy is the rejection of the transforming virus by the immune system.

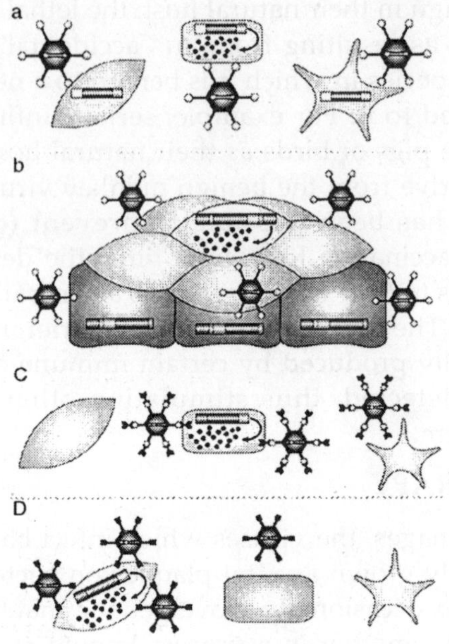

Fig. Viral Therapy

Oncolytic viruses are viruses that preferably infect cancer cells. While early efforts to employ these viruses in the therapy of cancer failed.

Chapter 7

Parasites

Parasitology is the study of parasites, their hosts, and the relationship between them. As a biological discipline, the scope of parasitology is not determined by the organism or environment in question, but by their way of life. This means it forms a synthesis of other disciplines, and draws on techniques from fields such as cell biology, bioinformatics, biochemistry, molecular biology, immunology, genetics, evolution and ecology. Parasites are organisms that obtain food and shelter by living on or within another organism. The parasite derives all benefits from association and the host may either not be harmed or may suffer the consequences of this association, a parasite disease.

The parasite is termed obligate when it can live only in association with a host or it is classified as facultative when it can live both in or on a host as well as in a free form. Parasites which live inside the body are termed endoparasites whereas those which exist on the body surface are called ectoparasites. Parasites that cause harm to the host are pathogenic parasites while those that benefit from the host without causing it any harm are known as commensals.

A parasite is an organism that obtains food and shelter from another organism and derives all benefits from this association. The parasite is termed obligate when it can live only in a host; it is classified as facultative when it can live both in a host as well as in free form. Parasites that live inside the body are termed endoparasites whereas those that exist on the body surface are called ecto-parasites. Parasites that cause harm to the host are pathogenic parasites while those that benefit from the host without causing it any harm are known as commensals. The organism that harbors the parasite and suffers a loss caused by the parasite is a host. The host in which the parasite lives its adult and sexual stage is the definitive host whereas the host in which a parasite lives as the larval and asexual stage is the intermediate host.

Other hosts that harbor the parasite and thus ensure continuity of the parasite's life cycle and act as additional sources of human infection are known as reservoir hosts. An organism (usually an insect) that is responsible for transmitting the parasitic infection is known as the vector.

TYPES OF PARASITE

Blood protozoa of major clinical significance include members of genera:

- Trypanosoma (*T. brucei* and *T. cruzi*);
- Leishmania (*L. donovani, L. tropica* and *L. braziliensis*);
- Plasmodium (*P. falciparum, P. ovale, P. malariae*)
- *Toxoplasma gondii*;
- Babesia (*B. microti*).

BABESIOSIS

ETIOLOGY

Babesia microti is the only member of the genus that infects man.

Morphology

The trophozoite is very similar to the ring form of the *Plasmodium* species (21A and B).

Life Cycle

The organism (sporozoite) is transmitted by a tick and enters the red cell where it undergoes mitosis and the organisms (merozoite) are released to infect other red cells. Ticks acquire the organism during feeding on an infected individual. In the tick, the organism divides sexually in the gut and migrates into the salivary gland (21C).

Symptoms

Babesiosis is associated with hemolytic anemia, jaundice, fever and hepatomegaly, usually 1-2 weeks after infection.

Diagnosis

Diagnosis is based on symptoms, patient history and detection of intraerythrocytic parasite in the patient or transfer of blood in normal hamsters which can be heavily parasitized.

Treatment and Control

Drugs of choice are clindamycin combined with quinine. The patient may recover spontaneously. One should avoid tick exposure and, if bitten, remove the tick from the skin immediately.

TRYPANOSOMIASIS

Epidemiology

T. b. gambiense is predominant in the western and central regions of Africa, whereas *T. b. rhodesiense* is restricted to the eastern third of the continent (2E). 6,000 to 10,000 human cases are documented annually. 35 million people and 25 million cattle are at risk. Regional epidemics of the disease are cause of major health and economic disasters.

Etiology

There are two clinical forms of African trypanosomiasis:

- A slowly developing disease caused by *Trypanosoma brucei gambiense* and
- A rapidly progressing disease caused by *T. brucei rhodesiense*.

Morphology

T. b. gambiense and *T. b. rhodesiense* are similar in appearance: The organism measures 10-30 micrometers x 1-3 micrometers. It has a single central nucleus and a single flagellum originating at the kinetoplast and joined to the body by an undulating membrane (2A-D). The outer surface of the organism is densely coated with a layer of glycoprotein, the variable surface glycoprotein (VSG).

Symptoms

The clinical features of Gambian and Rhodesian disease are the same, however they vary in severity and duration. Rhodesian disease progresses more rapidly and the symptoms are often more pronounced. The symptoms of the two diseases are also more pronounced in Caucasians than in the local African population. Classically, the progression of African trypanosomiasis can be divided into three stages: the bite reaction (chancre), parasitemia (blood and lymphoid tissues), and CNS stage.

Bite reaction: A non-pustular, painful, itchy chancre (4 A and B) forms 1-3 weeks after the bite and lasts 1-2 weeks. It leaves no scar.

Parasitemia: Parasitemia and lymph node invasion is marked by attacks of fever which starts 2-3 weeks after the bite and is accompanied by malaise, lassitude, insomnia headache and lymphadenopathy and edema. Painful sensitivity of palms and ulnar region to pressure (Kerandel's sign) may develop in some Caucasians. Very characteristic of Gambian disease is visible enlargement of the glands of the posterior cervical region (Winterbottom's sign). Febrile episodes may last few months as in Rhodesian disease or several years as in Gambian disease. Parasitemia is more prominent during the acute stage than during the recurrence episodes.

CNS Stage: The late or CNS stage is marked by changes in character and personality. They include lack of interest and disinclination to work, avoidance of acquaintances, morose and melancholic attitude alternating with exaltation, mental retardation and lethargy, low and tremulous speech, tremors of tongue and limbs, slow and shuffling gait, altered reflexes, etc. Males become impotent. There is a slow progressive involvement of cardiac tissue. The later stages are characterized by drowsiness and uncontrollable urge to sleep. The terminal stage is marked by wasting and emaciation. Death results from coma, intercurrent infection or cardiac failure.

The clinical features of Rhodesian disease are similar but briefer and more acute. The acuteness and severity of disease do not allow typical sleeping sickness. Death is due to cardiac failure within 6-9 months.

Life Cycle

The infective, metacyclic form of the trypanosome is injected into the primary host during a bite by the vector, the tsetse fly. The organism transforms into a dividing trypanosomal (trypomastigote) blood form (1B) as it enters the draining lymphatic and blood stream. The trypanosomal form enters the vector during the blood meal and travels through the alimentary canal to the salivary gland where it proliferates as the crithidial form (epimastigote) and matures to infectious metacyclic forms (1B).

Trypomastigotes can traverse the walls of blood and lymph capillaries into the connective tissues and, at a later stage, cross the choroid plexus into the brain and cerebrospinal fluid. The organism can be transmitted through blood transfusion.

Pathology and Immunology

An exact pathogenesis of sleeping sickness is not known, although immune complexes and inflammation have been suspected to be the mechanism of damage to tissues. The immune response against the organism does help to eliminate the parasite but it is not protective, since the parasite has a unique ability of altering its antigens, the VSG.

Consequently, there is a cyclic fluctuation in the number of parasites in blood and lymphatic fluids and each wave of parasite represents a different antigenic variant. The parasite causes polyclonal expansion of B lymphocytes and plasma cells and an increase in total IgM concentration.

It stimulates the reticuloendothelial function. It also causes severe depression of cell mediated and humoral immunity to other antigens.

Diagnosis

Detection of parasite in the bloodstream, lymph secretions and enlarged lymph node aspirate provides a definitive diagnosis in early (acute) stages. The parasite in blood can be concentrated by centrifugation or by the use of anionic support media. Cerebrospinal fluid must always be examined for organisms. Immuno-serology (enzyme-linked immune assay, immunofluorescence) may be indicative but does not provide definite diagnosis.

Treatment and Control

The blood stage of African trypanosomiasis can be treated with reasonable success with Pentamidine isethionate or Suramin. These drugs have been reported also to be effective in prophylaxis although they may mask early infection and thus increase the risk of CNS disease. Cases with CNS involvement should be treated with Melarsoprol, an organic arsenic compound.

The most effective means of prevention is to avoid contact with tsetse flies. Vector eradication is impractical due to the vast area involved. Immunization has not been effective due to antigenic variation.

TOXOPLASMOSIS

Etiology

Toxoplasma gondii is the organism responsible for toxoplasmosis

Epidemiology

Toxoplasma has worldwide distribution and 20%-75% of the population is seropositive without any symptomatic episode. However, the infection poses a serious threat in immunosuppressed individuals and pregnant females.

Morphology

The intracellular parasites (tachyzoite) are 3x6 microns, pear-shaped organisms that are enclosed in a parasite membrane to form a cyst measuring 10-100 microns in size. Cysts in cat feces (oocysts) are 10-13 microns in diameter.

Life Cycle

The natural life cycle of *T. gondii* occurs in cats and small rodents, although the parasite can grow in the organs (brain, eye, skeletal muscle, etc.) of any mammal or birds ( ). Cats gets infected by ingestion of cysts in flesh. Decystation occurs in the small intestine, and the organisms penetrate the submucosal epithelial cells where they undergo several generations of mitosis, finally resulting in the development of micro- (male) and macro- (female) gametocytes.

Fertilized macro-gametocytes develop into oocysts that are discharged into the gut lumen and excreted. Oocysts sporulate in the warm environment and are infectious to a variety of animals including rodents and man. Sporozoites released from the oocyst in the small intestine penetrate the intestinal mucosa and find their way into macrophages where they divide very rapidly (hence the name tachyzoites) and form a cyst which may occupy the whole cell.

The infected cells ultimately burst and release the tachyzoites to enter other cells, including muscle and nerve cells, where they are protected from the host immune system and multiply slowly (bradyzoites). These cysts are infectious to carnivores (including man). Unless man is eaten by a cat, it is a dead-end host.

Symptoms

Although Toxoplasma infection is common, it rarely produces symptoms in normal individuals. Its serious consequences are limited to pregnant women and immunodeficient hosts. Congenital infections occur in about 1-5 per 1000 pregnancies of which 5-10% result in miscarriage and 8-10% result in serious brain and eye damage to the fetus. 10-13% of the babies will have visual handicaps. Although 58-70% of infected women will give birth to a normal offspring, a small proportion of babies will develop active retino-chorditis or mental retardation in childhood or young adulthood. In immunocompetent adults, toxoplasmosis, may produce flu-like symptoms, sometimes associated with lymphadenopathy. In immunocompromised individuals, infection results in generalized parasitemia involvement of brain, liver lung and other organs, and often death.

Immunology

Both humoral and cell mediated immune responses are stimulated in normal individuals. Cell-mediated immunity is protective and humoral response is of diagnostic value.

Diagnosis

Suspected toxoplasmosis can be confirmed by isolation of the organism from tonsil or lymph gland biopsy.

Treatment

Acute infections benefit from pyrimethamine or sulphadiazine. Spiramycin is a successful alternative. Pregnant women are advised to avoid cat litter and to handle uncooked and undercooked meat carefully.

PNEUMOCYSTIS PNEUMONIA

Pneumocystis jiroveci (formerly known as Pneumocystis carinii) Pneumocystis jiroveci was formerly thought to be a protozoan but is now known to be a fungus. It is included here because pneumocystis pneumonia is often described as an opportunistic parasitic disease.

Pneumocystis pneumonia is an infection of immunosuppressed individuals and is particularly seen in AIDS patients. The organism is pleomorphic, exhibiting, at various stages of its life cycle: 1-2 micron sporozoites, 4-5 micron trophozoites and 6-8 micron cysts. It spreads from person to person in cough droplets.

Infection in immunosuppressed individuals results in interstitial pneumonia characterized by thickened alveolar septum infiltrated with lymphocytes and plasma cells. Pneumonia is associated with fever, tachypnea, hypoxia, cyanosis and asphyxia. Diagnosis is based on isolation of organisms from affected lungs. Trimethoprim sulphamethoxazole is the treatment of choice.

LEISHMANIASIS

Etiology

Several species of Leishmania are pathogenic for man: *L. donovani* causes visceral leishmaniasis (Kala-azar, black disease, dumdum fever); *L. tropica* (*L. t. major, L. t. minor* and *L. ethiopica*) cause cutaneous leishmaniasis (oriental sore, Delhi ulcer, Aleppo, Delhi or Baghdad boil); and *L. braziliensis* (also, *L. mexicana* and *L. peruviana*) are etiologic agents of mucocutaneous leishmaniasis (espundia, Uta, chiclero ulcer).

Epidemiology

Leishmaniasis is prevalent world wide: ranging from south east Asia, Indo-Pakistan, Mediterranean, north and central Africa, and south and central America.

Morphology

Amastigote (leishmanial form) is oval and measures 2-5 microns by 1 - 3 microns (10A-D), whereas the leptomonad measures 14 - 20 microns by 1.5 - 4 microns, a similar size to trypanosomes.

Life Cycle

The organism is transmitted by the bite of several species of blood-feeding sand flies (*Phlebotomus*) which carry the promastigote in the anterior gut and pharynx. The parasites gain access to mononuclear phagocytes where they transform into amastigotes and divide until the infected cell ruptures. The released organisms infect other cells. The sandfly acquires the organisms during the blood meal; the amastigotes transform into flagellate promastigotes and multiply in the gut until the anterior gut and pharynx are packed. Dogs and rodents are common reservoirs.

Symptoms

Visceral leishmaniasis (kala-azar, dumdum fever): *L.*

donovani organisms in visceral leishmaniasis are rapidly eliminated from the site of infection, hence there is rarely a local lesion, although minute papules have been described in children. They are localized and multiply in the mononuclear phagocytic cells of spleen, liver, lymph nodes, bone marrow, intestinal mucosa and other organs.

One to four months after infection, there is occurrence of fever, with a daily rise to 102-104 degrees F, accompanied by chills and sweating. The spleen and liver progressively become enlarged (11B, C and E). With progression of the diseases, skin develops hyperpigmented granulomatous areas (kala-azar means black disease). Chronic disease renders patients susceptible to other infections. Untreated disease results in death.

Cutaneous leishmaniasis (Oriental sore, Delhi ulcer, Baghdad boil): In cutaneous leishmaniasis, the organism (*L. tropica*) multiplies locally, producing of a papule, 1-2 weeks (or as long as 1-2 months) after the bite. The papule gradually grows to form a relatively painless ulcer. The centre of the ulcer encrusts while satellite papules develop at the periphery. The ulcer heals in 2-10 months, even if untreated but leaves a disfiguring scar. The disease may disseminate in the case of depressed immune function.

Mucocutaneous leishmaniasis (espundia, Uta, chiclero): The initial symptoms of mucocutaneous leishmaniasis are the same as those of cutaneous leishmaniasis, except that in this disease the organism can metastasize and the lesions spread to mucoid (oral, pharyngeal and nasal) tissues and lead to their destruction and hence sever deformity (12E). The organisms responsible are *L. braziliensis*, *L. mexicana* and *L. peruviana*.

Pathology

Pathogenesis of leishmaniasis is due to an immune reaction to the organism, particularly cell mediated immunity. Laboratory examination reveals a marked leukopenia with relative monocytosis and lymphocytosis, anemia and thrombocytopenia. IgM and IgG levels are extremely elevated due to both specific antibodies and polyclonal activation.

Diagnosis

Diagnosis is based on a history of exposure to sandfies, symptoms and isolation of the organisms from the lesion aspirate or biopsy, by direct examination or culture. A skin test (delayed hypersensitivity: Montenegro test) and detection of anti-leishmanial antibodies by immuno-fluorescence are indicative of exposure.

Treatment and Control

Sodium stibogluconate (Pentostam) is the drug of choice. Pentamidine isethionate is used as an alternative. Control measures involve vector control and avoidance. Immunization has not been effective.

MALARIA

Etiology

Four *Plasmodium* species are responsible for human malaria These are *P. falciparum*, *P. vivax*, *P. ovale* and *P. malariae*.

Epidemiology

There are an estimated 200 million global cases of malaria leading a mortality of more than one million people per year. *P. falciparum* (malignant tertian malaria) and *P. malariae* (quartan malaria) are the most common species of malarial parasite and are found in Asia and Africa. *P. vivax* (benign tertian malaria) predominates in Latin America, India and Pakistan, whereas, *P. ovale* (ovale tertian malaria) is almost exclusively found in Africa (12G).

Morphology

Malarial parasite trophozoites are generally ring shaped, 1-2 microns in size, although other forms (ameboid and band) may also exist. The sexual forms of the parasite (gametocytes) are much larger and 7-14 microns in size. *P. falciparum* is the largest and is banana shaped while others are smaller and round. *P. vivax* causes stippling of infected red cells.

Life Cycle

Malarial parasites are transmitted by the infected female anopheline mosquito which injects sporozoites present in the saliva of the insect. Sporozoites infect the liver parenchymal cells where they may remain dormant (hypnozoites) or undergo stages of schizogony to produce schizonts and merogony to produce merozoites (meronts). When parenchymal cells rupture, thousands of meronts are released into blood and infect the red cells. *P. ovale* and *P. vivax* infect immature red blood cells whereas *P. malariae* infects mature red cells. *P. falciparum* infects both. In red cells, the parasites mature into trophozoites.

These trophozoites undergo schizogony and merogony in red cells which ultimately burst and release daughter merozoites. Some of the merozoites transform into male and female gametocytes while others enter red cells to continue the erythrocytic cycle. The gametocytes are ingested by the female mosquito, the female gametocyte transforms into ookinete, is fertilized, and forms an oocyst in the gut. The oocyte produces sporozoites (sporogony) which migrate to the salivary gland and are ready to infect another host. The liver (extraerythrocytic) cycle takes 5-15 days whereas the erythrocytic cycle takes 48 hours or 72 hours (*P. malariae*). Malaria can be transmitted by transfusion and transplacentally.

Symptoms

The symptomatology of malaria depends on the parasitemia, the presence of the organism in different organs and the parasite burden. The incubation period varies generally between 10-30 days. As the parasite load becomes significant, the patient develops headache, lassitude, vague pains in the bones and joints, chilly sensations and fever. As the disease progresses, the chills and fever become more prominent.

The chill and fever follow a cyclic pattern (paroxysm) with the symptomatic period lasting 8-12 hours. In between the symptomatic periods, there is a period of relative normalcy, the duration of which depends upon the species of the infecting parasite. This interval is about 34-36 hours in the case of *P. vivax* and *P. ovale* (tertian malaria), and 58-60 hours in the case of *P. malariae* (quartan malaria). Classical tertian paroxysm is rarely seen in *P. falciparum* and persistent spiking or a daily paroxysm is more usual.

The malarial paroxysm is most dramatic and frightening. It begins with a chilly sensation that progresses to teeth chattering, overtly shaking chill and peripheral vasoconstriction resulting in cyanotic lips and nails (cold stage). This lasts for about an hour. At the end of this period, the body temperature begins to climb and reaches 103-106 degrees F. Fever is associated with severe headache, nausea (vomiting) and convulsions. The patient experiences euphoria, and profuse perspiration and the temperature begins to drop. Within a few hours the patient feels exhausted but symptomless and remains symptomatic until the next paroxysm.

Each paroxysm is due to the rupture of infected erythrocytes and release of parasites. Without treatment, all species of human malaria may ultimately result in spontaneous cure except with *P. falciparum* which becomes more severe progressively and results in death. This organism causes sequestration of capillary vasculature in the brain, gastrointestinal and renal tissues. Chronic malaria results in splenomegaly, hepatomegaly and nephritic syndromes.

Pathology and Immunology

Symptoms of malaria are due to the release of massive number of merozoites into the circulation. Infection results in the production of antibodies which are effective in containing the parasite load. These antibodies are against merozoites and schizonts. The infection also results in the activation of the reticuloendothelial system (phagocytes). The activated macrophages help in the destruction of infected (modified) erythrocytes and antibody-coated merozoites. Cell mediated immunity also may develop and help in the elimination of infected erythrocytes. Malarial infection is associated with immunosuppression.

Diagnosis

Diagnosis is based on symptoms and detection of parasite in Giemsa stained blood smears. There are also antibody tests (20B).

Treatment and Control

Treatment is effective with various quinine derivatives (quinine sulphate, chloroquine, meflaquine and primaquine, etc.). Drug resistance, particularly in *P. falciparum* and to some extent in *P. vivax* is a major problem. Control measures are eradication of infected anopheline mosquitos. Vaccines are being developed and tried but none is available yet for routine use.

FACULTATIVE PARASITIC PROTOZOA

These are free-living amebae that occasionally cause serious human disease. They are of particular significance in immunocompromised hosts.

Negleria fowleri

This organism is a flagellate that may inhabit warm waters (spas, warm springs, heated swimming pools, etc.) and gain access via the nasal passage to the brain and cause encephalitis

Acanthemeba

Several species of free-living Acanthemeba are pathogenic to man. They normally reside in soil and can infect children who swallow dirt while playing on the ground. In normal individuals, the infection may cause mild disease (pharyngitis) or remain asymptomatic, but in immunodeficient individuals, the organism may penetrate the esophageal mucosa and reach the brain where it causes granulomatous encephalitis.

MOLECULAR PARASITOLOGY

The disease of trypanosomiasis and the life cycle of the parasites that cause it. The novel aspects of the life cycle lead us to ways in which the parasite has gained unusual biochemical pathways to cope with its niche. These offer potential sites for chemotherapy. Trypanosomes are successful parasites which manage to escape the host's immune response; this happens by a very complex mechanism of antigen switching and it is the knowledge of this mechanism that has led us to the first steps in developing an anti-trypanosome vaccine.

ENTEROBIUS VERMICULARIS

Epidemiology

Enterobiasis is by far the commonest helminthic infection in the US (18 million cases at any given time). The worldwide infection is about 210 million. It is an urban disease of children in crowded environment (schools, day care centers, etc.). Adults may get it from their children. The incidence in whites is much higher than in blacks.

Morphology

The female worm measures 8 mm x 0.5mm; the male is smaller. Eggs (60 micrometers x 27 micrometers) are ovoid but asymmetrically flat on one side.

Life Cycle

Infection occurs when embryonated eggs are ingested from the environment, with food or by hand to mouth contact. The embryonic larvae hatch in the duodenum and reach adolescence in jejunum and upper ilium. Adult worms descend into lower ilium, cecum and colon and live there for 7 to 8 weeks. The gravid females, containing more than 10,000 eggs migrate, at night, to the perianal region and deposit their eggs there. Eggs mature in an oxygenated, moist environment and are infectious 3 to 4 hours later. Man-to-man and auto infection are common. Man is the only host.

Symptoms

Enterobiasis is relatively innocuous and rarely produces serious lesions. The most common symptom is perianal, perineal and vaginal irritation caused by the female migration.

The itching results in insomnia and restlessness. In some cases gastrointestinal symptoms (pain, nausea, vomiting, etc.) may develop. The conscientious housewife's mental distress, guilt complex, and desire to conceal the infection from her friends and mother-in-law is perhaps the most important trauma of this persistent, pruritic parasite.

Diagnosis

Diagnosis is made by finding the adult worm or eggs in the perianal area, particularly at night. Scotch tape or a pinworm paddle is used to obtain eggs.

Treatment and Control

Two doses (10 mg/kg; maximum of 1g each) of Pyrental Pamoate two weeks apart gives a very high cure rate. Mebendazole is an alternative. The whole family should be treated, to avoid reinfection. Bedding and underclothing must be sanitized between the two treatment doses. Personal cleanliness provides the most effective in prevention.

TRYPANOSOMA CRUZI

CHAGAS DISEASE

Chagas Disease is carried by reduvid bugs including the assasin bugs and rhodnius which infect the patient when they defecate after taking a blood meal. The symptoms of Chagas' disease are: chronic infection, neurological disorders (including dementia), megacolon megaesophagus, and damage to the heart muscle.

Chagas' disease is often fatal unless treated. In acute disease, there is often severe anemia, muscle pain and neurological disorders. The latter are common in children under 2 years in which death may occur in about a month.

Chronic disease may be mild and sometimes asymptomatic but there may be damage to nerves causing cessation of gut muscle contractions, irregular heartbeat and destruction of nervous system motor centers. The chronic form of the disease is found in adults but most likely arises from a childhood infection. *T. cruzi* can cross the placenta and so chronically-infected mothers can infect their babies which may succumb to the very acute form of the disease.

THE AFRICAN TRYPANOSOMES

T. cruzi escapes from host immune response but it does not do so by changing its antigenic coat. Thus, there is no antigenic variation. Instead, it escapes by hiding inside cells. The disease starts after a bite by the insect vector. After 7-14 days the trypanosomes arrive in the lymph nodes where they divide. Here, they form aggregates called pseudocysts..

When the pseudocysts rupture, the released parasites can enter cells in various parts of the body including lymphatic tissue, muscle and tissue around nerve ganglia. The invasion of the cardiac nerve ganglia is the cause of much of the heart disease in areas where *T. Cruzi* is found.

Infection of the Host Cell

Amastigotes bind to the cell surface (e.g. a monocyte) via a variety of receptor proteins including fibronectin. Sialic acid is important as cells deficient in sialic acid are not penetrated. Interestingly, *T. cruzi* has an enzyme called trans-sialidase which actually puts more sialic acid onto cells, thereby enhancing uptake.

Uptake is by a process of induced phagocytosis. Lysosomes migrate to the cell surface when cells come in contact with *T. cruzi* and the parasite enters a cytoplasmic vacuole, the parasitophorous vacuole, that is formed from a lysosome.

Thus, agents that induce the migration of lysosomes from peri-nuclear areas so that they underlie the plasma membrane, enhance infectivity. Conversely, blocking lysosomal function stops infection.

Somehow, the parasites escape from the destructive potential of the lysosome and after about an hour *T. cruzi* releases a protein toxin that inserts into the membrane at the low pH typical of this organelle. As a result *T. cruzi* escapes into the cytoplasm as the lysosomal membrane is destroyed.

NEMATODES

INTESTINAL HELMINTHS

Intestinal nematodes of importance to man are Ascaris lumbricoides (roundworm), Trichinella spiralis (trichinosis), Trichuris trichiura (whipworm), Enterobius vermicularis (pinworm), Strongyloides stercoralis (Cochin-china diarrhea), Ancylostoma duodenale and Necator americanes (hook-worms) and Dracunculus medinensis (fiery serpents of the Israelites). E.vermicularis and T. trichiura are exclusively intestinal parasites. Other helminths listed above have both intestinal and tissue phases.

Ascaris Lumbricoides (Large Intestinal Roundworm)

Epidemiology

The annual global morbidity due to ascaris infections is estimated at 1 billion with a mortality of 20,000. Ascariasis can occur at all ages, but it is more prevalent in the 5 to 9 years age group. The incidence is higher in poor rural populations.

Morphology

The average female worm measures 30 cm x 5 mm. The male is smaller.

Life Cycle

The infection occurs by ingestionof food contaminated with infective eggs which hatch in the upper small intestine. The larvae (250 x 15 micrometers) penetrate the intestinal wall and enter the venules or lymphatics. The larvae pass through the liver, heart and lung to reach alveoli in 1 to 7 days during which period they grow to 1.5 cm. They migrate up the bronchi, ascend the trachea to the glottis, and pass down the esophagus to the small intestine where they mature in 2 to 3 months. A female may live in the intestine for 12 to 18 months and has a capacity of producing 25 million eggs at an average daily output of 200,000.

The eggs are excreted in feces, and under suitable

conditions (21 to 30 degrees C, moist, aerated environment) infective larvae are formed within the egg. The eggs are resistant to chemical disinfectant and survive for months in sewage, but are killed by heat (40 degrees C for 15 hours). The infection is man to man. Auto infection can occur.

Symptoms

Symptoms are related to the worm burden. Ten to twenty worms may go unnoticed except in a routine stool examination. The commonest complaint is vague abdominal pain. In more severe cases, the patient may experience listlessness, weight loss, anorexia, distended abdomen, intermittent loose stool and occasional vomiting. During the pulmonary stage, there may be a brief period of cough, wheezing, dyspnea and sub-sternal discomfort. Most symptoms are due to the physical presence of the worm.

Diagnosis

Diagnosis is based on identification of eggs (40 to 70 micrometers by 35 to 50 micrometers - 2) in the stool.

Treatment and Prevention

Mebendazole, 200 mg, for adults and 100 mg for children, for 3 days is effective. Good hygiene is the best preventive measure.

TRICHINELLA SPIRALIS

Epidemiology

Trichinosis is related to the quality of pork and consumption of poorly cooked meat. Autopsy surveys indicate about 2 percent of the population is infected. The mortality rate is low.

Morphology

The adult female measures 3.5 mm x 60 micrometers. The larvae in the tissue (100 micrometers x 5 micrometers) are coiled in a lemon-shaped capsule.

Life Cycle

Infection occurs by ingestion of larvae, in poorly cooked meat, which immediately invade intestinal mucosa and sexually differentiate within 18 to 24 hours. The female, after fertilization, burrows deeply in the small intestinal mucosa, whereas the male is dislodged (intestinal stage). On about the 5th day eggs begin to hatch in the female worm and young larvae are deposited in the mucosa from where they reach the lymphatics, lymph nodes and the blood stream (larval migration).

Larval dispersion occurs 4 to 16 weeks after infection. The larvae are deposited in muscle fiber and, in striated muscle, they form a capsule which calcifies to form a cyst. In non-striated tissue, such as heart and brain, the larvae do not calcify; they die and disintegrate. The cyst may persist for several years. One female worm produces approximately 1500 larvae. Man is the terminal host. The reservoir includes most carnivorous and omnivorous animals.

Symptoms

Trichinosis symptoms depend on the severity of infection: mild infections may be asymptomatic. A larger bolus of infection produces symptoms according to the severity and stage of infection and organs involved.

Pathology and Immunology

Trichinella pathogenesis is due the presence of large numbers of larvae in vital muscles and host reaction to larval metabolites. The muscle fibers become enlarged edematous and deformed. The paralyzed muscles are infiltrated with neutrophil, eosinophils and lymphocytes. Splenomegaly is dependent on the degree of infection. The worm induces a strong IgE response which, in association with eosinophils, contributes to parasite death.

Diagnosis

Diagnosis is based on symptoms, recent history of eating raw or undercooked meat and laboratory findings (eosinophilia, increased serum creatine phosphokinase and lactate dehydrogenase and antibodies to *T. spiralis*).

Treatment and Control

Steroids are use for treatment of inflammatory symptoms and Mebendazole is used to eliminate worms. Elimination of parasite infection in hogs and adequate cooking of meat are the best ways of avoiding infection.

STRONGYLOIDES STERCORALIS

Epidemiology

Threadworm infection, also known as Cochin-China diarrhea, estimated at 50 to 100 million cases worldwide, is an infection of the tropical and subtropical areas with poor sanitation. In the United States, it is prevalent in the South and among Puerto Ricans.

Morphology

The size and shape of threadworm varies depending on whether it is parasitic or free-living. The parasitic female is larger (2.2 mm x 45 micrometers) than the free-living worm (1 mm x 60 micrometers). The eggs, when laid are 55 micrometers by 30 micrometers.

Life Cycle (sh)

The infective larvae of *S. stercoralis* penetrate the skin of man, enter the venous circulation and pass through the right heart to lungs, where they penetrate into the alveoli. From there, the adolescent parasites ascend to the glottis, are swallowed, and reach the upper part of the small intestine, where they develop into adults. Ovipositing females develop in 28 days from infection.

The eggs in the intestinal mucosa, hatch and develop into rhabditiform larvae in man. These larvae can penetrate through the mucosa and cycle back into the blood circulation, lung, glottis and duodenum and jejunum; thus they continue the auto infection cycle. Alternatively, they are passed in the feces, develop into infective filariform larvae and enter another host to complete the direct cycle. If no suitable host is found, the larvae mature into free-living worm and lay eggs in the soil. The eggs hatch in the soil and produce rhabditiform larvae which develope into infective filariform larvae and enter a new host (indirect cycle), or mature into adult worms to repeat the free-living cycle.

Symptoms

Light infections are asymptomatic. Skin penetration causes itching and red blotches. During migration, the organisms cause bronchial verminous pneumonia and, in the duodenum, they cause a burning mid-epigastric pain and tenderness accompanied by nausea and vomiting. Diarrhea and constipation may alternate. Heavy, chronic infections result in anemia, weight loss and chronic bloody dysentery. Secondary bacterial infection of damaged mucosa may produce serious complications.

Diagnosis

The presence of free rhabditiform larvae in the feces is diagnostic. Culture of stool for 24 hours will produce filariform larvae.

Treatment and Control

Ivermectin or thiabendozole can be used effectively. Direct and indirect infections are controlled by improved hygiene and auto-infection is controlled by chemotherapy.

ANCYLOSTOMA DUODENALE

Epidemiology

Hookworms parasitize more than 900 million people worldwide and cause daily blood loss of 7 million liters. Ancylostomiasis is the most prevalent hookworm infection and is second only to ascariasis in infections by parasitic worms. *N. americanes* (new world hookworm) is most common in the Americas, central and southern Africa, southern Asia,

Indonesia, Australia and Pacific Islands. *A. duodenale* (old world hookworm) is the dominant species in the Mediterranean region and northern Asia.

Morphology

Adult female hookworms are about 11 mm x 50 micrometers. Males are smaller. The anterior end of *N. americanes* is armed with a pair of curved cutting plates whereas *A. duodenale* is equipped with one or more pairs of teeth. Hookworm eggs are 60 micrometers x 35 micrometers.

Life Cycle (sh)

The life cycle of hookworms is identical to that of threadworms, except that hookworms are not capable of a free-living or auto-infectious cycle. Furthermore, *A. duodenale* can infect also by oral route.

Symptoms

Symptoms of hookworm infection depend on the site at which the worm is present and the burden of worms. Light infection may not be noticed.

Diagnosis

Diagnosis is made by identification of hookworm eggs in fresh or preserved feces. Species of hookworms cannot be distinguished by egg morphology.

Treatment and Control

Mebendazole, 200 mg, for adults and 100 mg for children, for 3 days is effective. Sanitation is the chief method of control: sanitary disposal of fecal material and avoidance of contact with infected fecal material.

DRACUNCULUS MEDINENSIS

Epidemiology

Guinea worm is estimated to infect about 50 million people in North, West and Central Africa, southwestern Asia, the West Indies and northeastern South America.

Morphology

The adult female worm measures 50-120 cm by 1 mm and the male is half that size.

Life Cycle

The infection is caused by ingestion of water contaminated with water fleas (Cyclops) infected with larvae. The rhabtidiform larvae penetrate the human digestive tract wall, lodge in the loose connective tissues and mature into the adult form in 10 to 12 weeks. In about a year, the gravid female migrates to the subcutaneous tissue of organs that normally come in contact with water and discharges its larvae into the water. The larvae are picked up by Cyclops, in which they develop into infective form in 2 to 3 weeks.

Symptoms

If the worm does not reach the skin, it dies and causes little reaction. In superficial tissue, it liberates a toxic substance that produces a local inflammatory reaction in the form of a sterile blister with serous exudation.

The worm lies in a subcutaneous tunnel with its posterior end beneath the blister, which contains clear yellow fluid. The course of the tunnel is marked with induration and edema. Contamination of the blister produces abscesses, cellulitis, extensive ulceration and necrosis.

Diagnosis

Diagnosis is made from the local blister, worm or larvae. The outline of the worm under the skin may be revealed by reflected light.

Treatment

Treatment includes the extraction of the adult guinea worm by rolling it a few centimeters per day or preferably by multiple surgical incisions under local anaesthesia. Metronidazole is effective in killing the worm. Protection of drinking water from being contaminated with Cyclops and larvae are effective preventive measures.

TRICHURIS TRICHIURA

Epidemiology

Trichuriasis is a tropical disease of children (5 to 15 yrs) in rural Asia (65% of the 500-700 million cases). It is, however, seen in the two Americas, mostly in the South and is concentrated in families and groups with poorer sanitary habits.

Morphology

The female organism is 50 mm long with a slender anterior (100 micrometer dia,eter) and a thicker (500 micrometers diameter) posterior end. The male is smaller and has a coiled posterior end. The Trichuris eggs are lemon or football shaped and have terminal plugs at both ends.

Life Cycle

Infection occurs by ingestion of embryonated eggs in soil. The larva escapes the shell in the upper small intestine and penetrates the villus where it remains for 3 to 10 days. Upon reaching adolescence, the larvae pass to the cecum and embed in the mucosa. They reach the ovipositing age in 30 to 90 days from infection, produce 3000 to 10,000 eggs per day and may live as long as 5 to 6 years. Eggs passed in feces embryonate in moist soil within 2 to 3 weeks. The eggs are less resistant to desiccation, heat and cold than ascaris eggs. The embryo is killed under desiccation at 37 degrees C within 15 minutes. Temperatures of 52 degrees C and -9 degrees C are lethal.

Symptoms

Symptoms are determined largely by the worm burden: less than 10 worms are asymptomatic. Heavier infections (e.g., massive infantile trichuriasis) are characterized by chronic profuse mucus and bloody diarrhea with abdominal pains and edematous prolapsed rectum. The infection may result in malnutrition, weight loss and anemia and sometimes death.

Diagnosis

Diagnosis is based on symptoms and the presence of eggs in feces (50 to 55 x 20 to 25 micrometers).

Treatment and Control

Mebendazole, 200 mg, for adults and 100 mg for children, for 3 days is effective. Accompanying infections must be treated accordingly. Improved hygiene and sanitary eating habits are most effective in control.

ANCYLOSTOMA BRAZILIENSIS

Creeping eruption is prevalent in many tropical and subtropical countries and in the US especially along the Gulf and southern Atlantic states. The organism is primarily a hookworm of dogs and cats but the filariform larvae in animal feces can infect man and cause skin eruptions. Since the larvae have a tendency to move around, the eruption migrates in the skin around the site of infection.

The symptoms last the duration of larval persistence which ranges from 2 to 10 weeks. Light infection can be treated by freezing the involved area. Heavier infections are treated with Mebendazole. Infection can be avoided by keeping away from water and soil contaminated with infected feces.

TOXOCARA CANIS AND T. CATTI

These are roundworms of dogs and cats but they can infect humans and cause damage of the visceral organs. Eggs from feces of infected animals are swallowed by man and hatch in the intestine. The larvae penetrate the mucosa, enter the circulation and are carried to liver, lungs, eyes and other organs where they cause inflammatory necrosis. Symptoms are due to the inflammatory reaction at the site of infection. The most serious consequence of infection may be loss of sight if the worm localizes in the eye. Treatment includes Mebendazole to eliminate the worm and prednisone for inflammatory symptoms. Avoidance of infected dogs and cats is the best prevention.

BLOOD AND TISSUE HELMINTHS

The major blood and tissue parasites of man are microfilaria. These include:

- *Loa loa* (eye worm).

- *Wuchereria bancrofti*
- *Onchocerca volvulus*

LOA LOA

Loasis is limited to the areas of African equatorial rain forest. The incidence in endemic areas varies greatly (8 to 75 percent). The larger, female organisms are 60 mm by 500 micrometers; males are 35mm by 300 micrometers in size.

The circulating microfilaria are 300 micrometers by 7 micrometers; the infective larvae in the fly are 200 micrometers by 30 micrometers. The life cycle of *Loa loa* is identical to that of onchocerca except that the vector for this worm is the deer fly. The infection results in subcutaneous (Calabar) swelling, measuring 5 to 10 cm in diameter, marked by erythema and angioedema, usually in the extremities.

The organism migrates under the skin at a rate of up to an inch every two minutes. Consequently, the swelling appears spontaneously, persists for 4 to 7 days and disappears, and is known as fugitive or Calabar swelling.

The worm usually causes no serious problems, except when passing through the orbital conjunctiva or the nose bridge. The diagnosis is based on symptoms, history of deer fly bite and presence of eosinophilia. Recovery of worms from the conjunctiva is confirmatory. Treatment and control are the same as those for onchocerciasis.

ONCHOCERCA VOLVULUS

Epidemiology

Onchocerciasis is prevalent throughout eastern, central and western Africa, where it is the major cause of blindness. In the Americas, it is found in Guatemala, Mexico, Colombia and Venezuela. The disease is confined to neighborhoods of low elevation with rapidly flowing small streams where black flies breed. Man is the only host.

Morphology

Adult female onchocerca measure 50 cm by 300

micrometers, male worms are much smaller. Infective larvae of *O. volvulus* are 500 micrometers by 25 micrometers.

Life Cycle

Infective larvae are injected into human skin by the female black fly (*Simulium damnosum*) where they develop into adult worms in 8 to 10 months. The adults usually occur as group of tightly coiled worms (2 to 3 females and 1 to 2 males). The gravid female releases microfilarial larvae, which are usually distributed in the skin. They are picked up by the black fly during a blood meal. The larvae migrate from the gut of the black fly to the thoracic muscle where they develop into infective larvae in 6 to 8 days. These larvae migrate to the head of the fly and then are transmitted to a second host.

Symptoms

Onchocerciasis results in nodular and erythematous lesions in the skin and subcutaneous tissue due to a chronic inflammatory response to persistent worm infection. During the incubation period of 10 to 12 months, there is eosinophilia and urticaria. Ocular involvement consists of trapping of microfilaria in the cornea, choroid, iris and anterior chambers, leading to photophobia, lacrimation and blindness.

Diagnosis

Diagnosis is based on symptoms, history of exposure to black flies and presence of microfilaria in nodules.

Treatment and Control

Diethylcarbamazine is effective in killing the worm. Destruction of microfilaria produces extreme allergic reaction which can be controlled with corticosteroids. Prevention measures include vector control, treatment of infected individuals and avoidance of black fly.

WUCHERERIA BANCROFTI

Epidemiology

W. bancrofti is strictly a human pathogen and is distributed in tropical areas worldwide, whereas *B. malayi* infects a number of wild and domestic animals and is restricted to South-East Asia. Mosquitoes are vectors for both parasites.

Morphology

These two organisms are very similar in morphology and in the diseases they cause. Adult female *W. bancrofti* found in lymph nodes and lymphatic channels are 10 cm x 250 micrometers whereas males are only half that size. Microfilaria found in blood are only 260 micrometers x 10 micrometers. Adult *B. malayi* are only half the size of *W. bancrofti* but their microfilaria are only slightly smaller than *W. bancrofti*.

Life Cycle

Filariform larvae enter the human body during a mosquito bite and migrate to various tissues. There, they may take up to a year to mature and produce microfilaria which migrate to lymphatics and, at night, enter the blood circulation. Mosquitos are infected during a blood meal. The microfilaria grow 4 to 5 fold in the mosquito in 10 to 14 days and become infective for man.

Symptoms

Symptoms include lymphadenitis and recurrent high fever every 8 to 10 weeks, which lasts 3 to 7 days. There is progressive lymphadenitis due to an inflammatory response to the parasite lodged in the lymphatic channels and tissues. As the worm dies, the reaction continues and produces a fibro-proliferative granuloma which obstructs lymph channels and causes lymphedema and elephantiasis. The stretched skin is susceptible to traumatic injury and infections. Microfilaria cause eosinophilia and some splenomegaly. Not all infections lead to elephantiasis. Prognosis, in the absence of elephantiasis, is good.

Diagnosis

Diagnosis is based on history of mosquito bites in endemic areas, clinical findings and presence of microfilaria in blood samples collected at night.

Treatment and Control

Diethylcarbamazine quickly kills the adults worms or sterilizes the female. It is given 2 mg/kg orally for 14 days. Steroids help alleviate inflammatory symptoms. Cooler climate reduces the inflammatory reaction.

TAPEWORM

DWARF TAPEWORM

Hymenolepis nana

This is a small tapeworm (20 x 0.7 mm) which infects children. Rodents are the reservoir. Infection is by the oro-fecal mode and, hence, cross infection and auto infection by eggs in feces in normal. The worm develops from ingested eggs into an adult in the small intestine and resides there for several weeks. Light infections produce vague abdominal disturbances but heavier infections may cause enteritis. Diagnosis is based on finding eggs in the feces. Nicolsamide is the drug of choice. Hygiene is the best control.

ECHINOCOCCUS GRANULOSUS

Epidemiology

The organism is common in Asia, Australia, Eastern Africa, southern Spain, southern parts of South America and northern parts of North America. The incidence of human infection about 1 to 2 per 1000 population and may be higher in rural areas of affected regions.

Life Cycle

The adult worm lives in domestic and wild carnivorous animals. Eggs, passed by infected animals, are ingested by the grazing farm animals or man, localize in different organs and develop into hydatid cysts containing many larvae (proto-scolices or hydatid sand). When other animals consume infected organs of these animals, proto-scolices escape the cyst, enter the small intestine and develop into adult worms.

Echinococcus eggs, when swallowed by man, produce

embryos that penetrate the small intestine, enter the circulation and form cysts in liver, lung, bones, and sometimes, brain. The cyst is round and measures 1 to 7 cm in diameter, although it may grow to be 30 cm. The cyst consists of an outer anuclear hyaline cuticula and an inner nucleated germinal layer containing clear yellow fluid. Daughter cysts attach to the germinal layer, although some cysts, known as brood cysts, may have only larvae (hydatid sand). Man is a dead end host.

Symptoms

The symptoms, comparable to those of a slowly growing tumor, depend upon the location of the cyst. Large abdominal cysts produce increasing discomfort. Liver cysts cause obstructive jaundice. Peribronchial cysts may produce pulmonary abscesses. Brain cysts produce intracranial pressure and Jacksonian epilepsy. Kidney cysts cause renal dysfunction. The contents of a cyst may produce anaphylactic responses.

Diagnosis

Clinical symptoms of a slow-growing tumor accompanied by eosinophilia are suggestive. Intradermal (Casoni) test with hydatid fluid is useful. Pulmonary cysts and calcified cysts can be visualized using x-rays. Antibodies against hydatid fluid antigens have been detected in a sizable population of infected individuals by ELISA or indirect hemagglutination test.

Treatment and Control

Treatment involves surgical removal of cyst or inactivation of hydatid sand by injecting the cyst with 10% formalin and its removal within five minutes. It has been claimed that a high dose of Mebendazole results in some success. Preventive measures involve avoiding contact with infected dogs and cats and elimination of their infection.

TAPE WORMS

CESTODES

Clinically important cestodes pathogenic to man are:

- Tenia solium (pork tapeworm),
- T. saginata (beef tapeworm),
- Diphyllobothrium lattum (fish or broad tapeworm),
- Hymenolepis nana (dwarf tapeworm)
- Echinococcus granulosus
- E. multilocularis (hydatid).

DIPHYLLOBOTHRIUM LATUM

EPIDEMIOLOGY

Fish tapeworm infection is distributed worldwide, in the subarctic and temperate regions; it is associated with eating of raw or improperly cooked fresh water fish.

Morphology

This is the longest tapeworm found in man, ranging from 3-10 meters with more than 3000 proglottids. The scolex resembles two almond-shaped leaves and the proglottids are broader than they are long, a morphology reflected in the organism's name. Eggs are 30 x 50 micrometers in size and contain an embryo with 3 pairs of hooklets.

Life Cycle

Man and other animals are infected by eating uncooked fish that contains plerocercoid larvae (15 x 2 mm) which attach to the small intestinal wall and mature into adult worms in 3 to 5 weeks. Eggs discharged from gravid proglottids in the small intestine are passed in the feces.

The egg hatches in fresh water to produce a ciliated coracidium which needs to be ingested by a water flea (Cyclops) where it develops into a procercoid larva. When infected Cyclops are ingested by the freshwater fish, the procercoid larva penetrates the intestinal wall and develops into a plerocercoid larva, infectious to man.

Symptoms

Clinical symptoms may be mild, depending on the number of worms. They include abdominal discomfort, loss of weight, loss of appetite and some malnutrition. Anemia and neurological problems associated with vitamin B12 deficiency are seen in heavily infected individuals.

Diagnosis

Diagnosis is based on finding many typical eggs and empty proglottids in feces. A history of raw fish consumption and residence in an endemic locality is helpful.

Treatment and Control

Praziquantel is the drug of choice. Freezing for 24 hours, thorough cooking or pickling of fish kills the larvae. Fish reservoirs should be kept free of raw sewage.

TENIA SOLIUM

Epidemiology

These cestodes have a worldwide distribution but incidence is higher in developing countries. Infection rate is as low as 1 per 1000 in most of North America and as high as 10% in the third world. Pork tapeworm shows a higher incidence but this is dependent on dietary habits.

Morphology

T. saginata can be up to 4 to 6 meters long and 12 mm broad; it has a pear-shaped head (scolex) with four suckers but no hooks or neck. It has a long flat body with several hundred segments (proglottids).

Each segment is about 18 x 6 mm with a branched uterus (15-30 branches). The egg is 35 x 45 micrometers, roundish and yellow-brown. It has peripheral radial striations and contains an embryo with 3 hooklets.

T. solium is slightly smaller than *T. saginata*. It has a globular scolex with four suckers and a circular row of hooks (rostellum) that gives it a solar appearance. There is a neck and it has a long flat body (0.1 meter in length). The proglottids are 5 x 10 mm with a 7-12 branch uterus. The eggs of *T. solium* and *T. saginata* are indistinguishable.

Life Cycle

A tapeworm larval cyst (cysticercus) is ingested with poorly cooked infected meat; the larva escapes the cyst and passes to the small intestine where it attaches to the mucosa by the scolex suckers. The proglottids develop as the worm matures in 3 to 4 months. The adult may live in the small intestine as long as 25 years and pass gravid proglottids with the feces. Eggs extruded from the proglottid contaminate and persist on vegetation for several days and are consumed by cattle or pigs in which they hatch and form cysticerci.

Symptoms

Light infections remain asymptomatic, but heavier infections may produce abdominal discomfort, epigastric pain, vomiting and diarrhea.

Cysticercosis

T. solium eggs can also infect humans and cause cysticercosis (larval cysts in lung, liver, eye and brain) resulting in blindness and neurological disorders. The incidence of cerebral cysticercosis can be as high 1 per 1000 population and may account for up to 20% of neurological case in some countries (e.g., Mexico); cysticercosis ocular involvement occurs in about 2.5% of patients and muscular involvement is as high as 10% (India).

Pathology and Immunology

Gastrointestinal symptoms are due to the presence of the tape worm. Cysticercosis symptoms are a result of inflammatory/immune responses. Antibodies are produced in cysticercosis and are useful epidemiological tools.

Diagnosis

Diagnosis is based on the recovery of eggs or proglottids in stool or from the perianal area. Cysticercosis is confirmed by the presence of antibodies.

Treatment and Control

Praziquantel is the drug of choice. Expulsion of scolex

must be assured to assume a satisfactory treatment. A thorough inspection of beef and pork, adequate cooking or freezing of meat are effective precautions, since cysticerci do not survive temperatures below -10° C and above 50°C.

LUNG FLUKE

PARAGONIMUS WESTERMANI

Epidemiology

Lung fluke is most commonly encountered in parts of Asia, Africa and South America.

Morphology

It is a plump reddish brown oval worm measuring 10 by 4 mm. The ovum measures 85 by 55 micrometers.

Life Cycle

Lung fluke infects man (and domestic carnivores) when crabmeat infested with encysted metacercaria is consumed. The metacercaria reach the small intestine, exit their shell and bore their way, as young flukes, through the intestinal wall, through the thoracic diaphragm and penetrate the lung. There, they become enclosed in 1 to 2 cm cysts and reach maturity. The eggs are found in the sputum or, if swallowed, in the feces, 2 to 3 months after infection. The eggs, when introduced in fresh water produce a miracidia which penetrates the suitable snail. In the snail they develop into cercaria which break out in water and penetrate gills, muscle or viscera of fresh water crabs and become encysted in flesh as metacercaria.

Symptoms

The fluke provokes the development of a fibrous tissue capsule with bloody purulent material containing eggs. There is inflammatory infiltrate around the capsule. The symptoms include a dry cough, followed by production of blood stained rusty brown sputum. Pulmonary pain and pleurisy may develop. Worms may migrate to the brain where they lay eggs and cause a granulomatous abscess resulting in symptoms similar to epilepsy.

Diagnosis

Diagnosis is based on history and symptoms. Eggs are found in rust coloured sputum, often being examined for tuberculosis.

Treatment and Control

Praziquantel taken orally is quite effective. Adequate cooking of crustaceans is a preventive measure. Improved sanitary conditions have lowered the infection rate in endemic areas.

TREMATODES

The most significant trematodes from a clinical point of view are:

- Blood flukes,
- Schistosoma mansoni,

SCHISTOSOMIASIS

The three species of Schistosoma have different geographic distributions. *S. hematobium* is prevalent in Africa, *S. mansoni* is found in Africa and America and *S. japonicum* is common in the far east.

Epidemiology

Approximately 250 million people are infected with schistosomes and 600 million are at risk.

Morphology

Adult worms are 10 to 20 mm long; the male has an unusual lamelliform shape with marginal folds forming a canal in which the slender female worm resides. Unlike other trematodes, schistosomes have separate sexes.

Life Cycle

Man is infected by cercaria in fresh water by skin

penetration. The cercaria travel through the venous circulation to the heart, lungs and portal circulation. In about 3 weeks, they mature and reach the mesenteric (*S. japonicum* and *S. mansoni*) or the bladder (*S. hematobium*) vessels where they live and ovulate for the duration of the host's life.

Eggs germinate as they pass through the vessel wall into the intestine or bladder and are excreted in feces (*S. japonicum* and *S. mansoni*) or urine (*S. hematobium*). In fresh water, the larval miracidium hatches out of the egg and swims about until it finds an appropriate snail. After two generations of multiplication in the snail, the fork-tailed cercariae emerge into the water and infect another human.

Symptoms

Penetration of cercariae causes transient dermatitis (swimmers' itch). The symptoms of schistosomiasis are primarily due to a reaction against the eggs and include splenomegaly, lymphadenopathy and diarrhea. In the bladder, they produce granulomatous lesions, hematuria and sometimes urethral occlusion. Most bladder cancers in endemic areas are associated with chronic infection. In the intestine, they cause polyp formation which, in severe cases, may result in life threatening dysentery.

In the liver, the eggs cause periportal fibrosis and portal hypertension resulting in hepatomegaly, splenomegaly and ascites. A gross enlargement of the esophageal and gastric veins may result in their rupture. *S. japonicum* eggs are sometimes carried to the central nervous system and cause headache, disorientation, amnesia and coma. Eggs carried to the heart produce arteriolitis and fibrosis resulting in enlargement and failure of the right ventricle.

Pathology and Immunology

The 'swimmers' itch is due to physical damage to the skin by proteases and other toxic substances secreted by the cercaria. The host develops both type I and type IV hypersensitivity reactions to schistomal secretions and egg constituents. Embryonated eggs cause collagenase-mediated damage to the vascular endothelium. Host immune responses, both humoral and cell mediated, have been shown to be of some protective value. IgE and eosinophil mediated cytotoxicity has been suggested as a mechanism of killing the adult worm.

Diagnosis

Diagnosis is based on a history of residence in an endemic area, swimmers' itch and other symptoms. The eggs are very characteristic and confirm diagnosis. *S. hematobium* eggs in urine (55 to 65 by 110 to 170 micrometers) have an apical spine or knob. *S. mansoni* eggs in feces (45 to 70 by 115-175 micrometers) have a spine on the side. *S. japonicum* eggs are more round with a vague spine on the side.

Treatment and Control

Praziquantel is effective against all species. Contaminated water should be avoided. Control measures include sanitary disposal of sewage and destruction of snails. No vaccine is available.

GIANT INTESTINAL FLUKE

FASCIOLOPSIS BUSKI

Epidemiology

This is a parasite of central and southeast Asia.

Morphology

The elongate oval fluke is 2 to 7 cm long and lives in the small intestine of man.

Life Cycle

Man is infected by ingesting water chestnuts contaminated with metacercaria which find access to the small intestine, attach themselves to the mucosa and mature in 25 to 30 days. The fluke eggs are passed in the feces and hatch in fresh water producing miracidia which must penetrate a suitable snail within hours. The miracidia in the snail develop into cercaria and enter fresh water where they attach themselves to water plants (water chestnut) and encyst to become metacercaria.

Symptoms

Epigastric pain, nausea and diarrhea are experienced, especially in the morning. In heavier infections, generalized edema and ascites occur.

Pathology

The fluke attaches itself to the intestinal mucosa where inflammation, ulceration and abscesses occur.

Diagnosis

Diagnosis is based on clinical symptoms in endemic areas. Eggs in feces (75 to 100 by 130 to 150 micrometers) provide the final diagnosis.

Treatment and Control

Praziquantel has proven effective. Water chestnuts from contaminated waters should be avoided. Sewage should be treated before disposal.

CHINESE LIVER FLUKE

Epidemiology

This is a widespread parasite of man, dogs and cats in the southeast of Asia. It is extraordinarily common in China and is also found in Korea and Japan. Related flukes parasitizing European cats (*Opisthorchis felinus*) and dogs (*O. viverini*) infect humans in the endemic areas.

Morphology

These are spindloid flukes measuring about 16 by 4 mm. The eggs measure 29 by 16 micrometers.

Life Cycle

Man is infected by eating raw or improperly cooked fish

which carries the infective metacercaria in a cyst. The cyst is digested and the larval worm migrates up the bile duct to the liver where it matures into an adult. The eggs, deposited in the biliary duct, pass in the feces and find their way to fresh water. Upon ingestion by a suitable fresh water operculate snail, the egg hatches to produce a miracidium. The miracidium in the snail develops into cercaria which breaks out in water to penetrate under the scales of fish. In fish, the cercaria encysts in the muscle and forms metacercaria infectious to man.

Symptoms

The worm causes irritation of the bile ducts which become dilated and deviated. The liver may enlarge, become necrotic and tender and its function may be impaired. Modest infections results in indigestion, epigastric discomfort, weakness and loss of weight. Heavier infections produce anemia, liver enlargement, slight jaundice, edema, ascites and diarrhea.

Diagnosis

Diagnosis is based on symptoms and the presence of endemic infection in the area. Definitive diagnosis is dependent on finding the characteristic eggs in the feces or biliary drainage.

Treatment and Control

Praziquantel has proven of value. Fish should be cooked well before consumption. Sewage must be treated before disposal.

ARTHROPODS

Arthropods play a major role in human disease. Most of these are as vectors of different pathogens and, in other sections of this text, we have dealt with such disease vectors. There are also a number of arthropods that cause harm due their venom but these are not parasites. Here, we shall deal with arthropods that are parasitic to and cause disease in man.

For example, myiasis (burial of larvae in tissue) is an obligatory step in the life cycle of some flies and incidental for others. Species that cause myiasis in the Americas are *Cochliomyia* (Screw worm fly), *Calliphora, Oestrus, Sarcophaga, Gastrophilus,* etc.

Myiasis may be cutaneous, arterial, intestinal or urinary in normal tissue or in pre-existing wounds, some of which may result from other infections. Larvae can burrow through necrotic or healthy tissue using their mandibular hooks aided by proteolytic enzymes. They can cause mechanical damage and the affected area may be the site of a secondary infection. Cutaneous myiasis may require surgical removal of burrowed larvae. Eggs and maggots may be washed from hair, skin and wounds with soap and water. Urinary myiasis usually clears itself. Purgation with anti-helminths may be necessary for gastrointestinal myiasis.

MITES

Scabies mite (*Sarcoptes scabei*) is the cause of scabies and is distributed worldwide. Epidemics of the disease may occur for long periods but mites may be common at all times in very poor communities with inadequate washing facilities. The mite transmitted by contact burrows into the skin on the webbing side of fingers, later spreading to the wrists, elbows and the rest of the body. The buttocks, women's breasts and external genitalia may be involved.

The mite tunnels itself through the upper layer of the skin depositing eggs. Larvae escape the tunnel and wander on the skin and start new burrows and mature there to continue the cycle. Scabies itch is due to the sensitization of the patient to the mite and eggs and is characteristically nocturnal. Septic pustules may develop after scratching, if the hygiene is poor. Diagnosis is made by the characteristic rash and by smearing black ink on the skin and observing burrows when the ink is wiped away. Microscopic examination of a skin scraping shows the mites.

Treatment involves swabbing of the whole body from the neck down with 1% malathion or benzene hexachloride (crotamiton for infants). Topical steroids must not be used. If possible, the whole family should be treated. Contact with an infested person should be avoided. Clothes should be washed in hot water.

FLEAS

Most fleas are of clinical significance to man because they are vectors for other parasites. However, the jigger flea or chigoe (*Tunga penetrans*) is a serious pest in the tropical and subtropical regions of the Americas and Africa. Diagnosis of tungiasis is rare in North America. An epidemiological study in a traditional fishing village in Ceará State, north-eastern Brazil found about 51% of the population was infested.

Both sexes feed on blood. The female flea, after insemination, burrows itself in the skin of the toes and the sole of the foot. The female swells to the size of a pea, produces eggs and dies in the tissue. There is local reaction to the bite and the eggs and dead flea produce reaction. The infested tissue can get infected and gangrenous; auto-amputation is not uncommon. Treatments are symptomatic: infestation may be physically removed; secondary infections are treated appropriately. Shoes should be worn in infested areas.

LICE

Three types of sucking lice are important for human health: *Pediculus humanus capitis* (head louse), *P. humanus humanus* (body louse) and *Pthirus pubis* (crab louse). Lice spend all of their life on one very specific host and both male and female feed on blood and leave one host only to transfer to another.

Head Lice

In the developed world, 2 to 10% of children are infested with head lice. Light infestations cause moderate itching of the scalp due to sensitization to louse saliva. Heavy infestations may result in fever, aches and secondary infections. Diagnosis is based on finding live lice or empty egg shells (nits measuring 0.8 x 0.3 mm) attached to hair, often behind the ears.

Topical application of soothing lotions relieves irritation. Hair should be washed with shampoo containing 1% benzene hexachloride (Kwell). Application of a mixture of pyrentins (0.2%) and pipronyl butoxide (2%) or copper oleate may be as effective and less toxic than benzene hexachloride.

Body Lice

The body louse is similar to the head louse except that it is found on the body and clothes. Diagnosis is based on finding eggs or nits (eggs) in seams of clothing. Symptoms and treatment are the same as those for head lice.

Pubic Lice

Pubic lice, also known as crab lice, infest widely-spaced coarse hair in the pubic area in adults or eye lashes in children. Transmission in adults is usually by sexual contact. Diagnosis is based on finding lice or nits in the infested area; crab lice may be difficult to see at the base of the hair. The pubic area is treated in the same manner as the infested head. Nits and lice may be removed from eye lashes with forceps. Ointments with physostigmine (0.25%) or yellow mercury oxide are effective.

CHIGGERS

Chiggers (the larvae of red mites or harvest mites - 8) (Trombiculidae) are an important group of ectoparasites affecting humans. They attach to the skin in the ankles, waistline, armpits and perianal area after the host walks through a grassy environment.

Contrary to popular belief, these mite larvae do not feed on blood and do not burrow into the skin. They pierce the skin near a hair follicle and feed on partially digested skin cells using enzymes in the chigger's saliva. They then drop off the host. The host reacts to the mouth parts and saliva of the mite, however, and after a few hours an erythematous papule appears that is highly pruritic. The intensity of the eruption depends on the sensitivity of the host and may be followed by fever.

Treatment with a local anesthetic is useful. Insect

repellents (DEET) may be effective in avoiding chigger bites. In south Asia, chiggers are the vectors for scrub typhus (*Orientia tsutsugamushi*), a rickettsial disease that can (rarely) be life-threatening.

BOT FLY AND TUMBU FLY - MYIASIS

Myiasis is the parasitism of a vertebrate host by the larva of a dipteran fly. It frequently occurs in domestic and wild animals and therefore is important in veterinary medicine. Human cases are rare but occur especially in tropical countries. In the United States, the most common agents of myiasis are *Dermatobia hominis* (bot fly, berne) and *Cordylobia anthropophaga* (tumbu fly). The former occurs in central (particularly Mexico and Belize) and south America and some Caribbean islands such as Trinidad, while the latter occurs in tropical Africa. The larvae of both of these flies are obligate parasites of mammals.

The human bot fly has a very interesting life cycle. The female fly captures a mosquito and lays its eggs on its abdomen. When the released mosquito takes a blood meal (usually on the face, scalp or extremities), the body heat of the mammal causes the eggs to hatch into a first instar larva, which drops onto the skin of its host.

The larva either enters the skin via the puncture wound made by the mosquito or actively penetrates the skin where it takes up residence in sub-cutaneous tissue. Here, the larva matures through second and third instars over a period of one or two months. The larvae are difficult to remove because of backward facing barbs. When they are mature, the larvae emerge and fall to the ground where they pupate into the adult fly and the life cycle repeats. The adult fly only lives a few days and does not feed.

The initial symptom of the presence of a bot fly is a cutaneous nodule that contains one larva (although this is not initially visible). The nodule is often called a warble or a furuncle and myiasis is often referred to as a furunculous disease. Unlike a puncture wound of a mosquito, the infected nodule discharges blood or serum continually because the larva needs to keep the wound open in order to breath. There is often pruritus and sometimes intense periodic shooting pain that occurs when the larva moves or matures to another instar. As the larva grows, movement can often be seen below the surface of the serosanguineous effusion. There may also be local lymphadenopathy, and fever and malaise if a secondary bacterial infection develops.

Treatment entails removal of the larva, although some patients prefer to allow the larva to develop and emerge naturally. In the case of *Dermatobia hominis*, the Achilles heel is the need for the larva to breath air though the open wound. Cutting off the parasite's air supply is most often done by applying a thick layer (at least 5mm) of Vaseline, although nail polish, adhesive tape and even bacon have been reported to be used.

Because of the lack of oxygen, the larva will usually emerge and can be pulled the remaining way out with forceps. However, it often takes a day for the larva to come far enough out to be grabbed with forceps. Surgery is not usually necessary unless the larva dies in situ and cannot be removed.

Tumbu fly (*Cordylobia anthropophaga*) is found in sub-Saharan Africa and causes a furuncular myiasis similar to bot fly. The patient comes in contact with eggs on the ground or deposited on clothes. After hatching, the larva burrows into the skin. The eggs are killed by ironing clothes but for many people that may be impractical.

TICKS

Ticks are are blood-sucking, opportunistic parasites that can attach to the skin of a variety of vertebrate hosts. They have no segmentation and are dorso-ventrally flat with four pairs of legs.

Although all stages of the tick life cycle can suck blood, it is normally the adult tick that poses a problem for humans. Human tick-associated diseases are most common in the summer months when the likelihood of contact increases during outdoor activities, usually in wooded areas. Being bitten by a tick is often painless and the presence of the tick may not be detected for some time. Often the tick poses no problem for the human host other than an erythromatous papule and it drops off after engorging on blood. Sometimes, the site of attachment may itch and become painful. Secondary infections of the wound site may occur, often as a result of the mouthparts remaining attached after the tick is removed. Ticks can attach anywhere on the body but are frequently found at the hairline, around the ears, groin, armpits etc.

Whilst the bites of most ticks are inconsequential, they can carry a number of human disease agents including viruses, bacteria and protozoa. There are two families of ticks: the hard ticks (Ixodidae) and the soft ticks (Argasidae). The Ixodidae attach to their host over a prolonged period of time (several days) while the Argasidae feed rapidly and then drop off. Consequently, they are frequently undetected.

Although most tick-associated problems arise from disease-causing organisms carried by the ticks, in one case – tick paralysis – the problem arises directly from toxins in the tick's saliva.

SOFT TICKS CARRIERS

Tick-borne Relapsing Fever

Tick-borne relapsing fever is a rare disease (about 25 cases per year in the United States) and is caused by several spirochete bacterial species of the *Borelia* family. The transmission agents are soft ticks of the genus *Ornithodoros*. Soft ticks (family *Argasidae*) differ in many ways from the so-called hard ticks (family *Ixodidae*), but the most important is that they take brief meals from their host and then drop off. The bite is usually painless.

Thus, they are far less likely to be found than the hard ticks that stay attached while feeding for hours. In the wild, these ticks are found in nesting materials when not feeding on their animal host. All stages of the life cycle can take blood meals.

The individual Borrelia species that cause tick-borne relapsing fever are usually associated with specific Ornithodoros tick vectors. B. hermsii is transmitted to humans by Ornithodoros hermsi, B. parkerii is transmitted by Ornithodoros parkeri and B. turicatae is transmitted by Ornithodoros turicata. Each tick is associated with a preferred environment and hosts.

Ornithodoros hermsi is found at higher altitudes (1500 – 8000 feet) where it is associated usually with ground squirrels, tree squirrels and chipmunks. Ornithodoros parkeri occurs at lower elevations and inhabit caves and the burrows of ground squirrels, prairie dogs and burrowing owls. Ornithodoros turicata occurs in caves and ground squirrel, prairie dog or burrowing owls burrows in the plains regions of the Southwest United States.

Symptoms

Initially, the patient experiences arthralgia, myalgia, headache, chills and fever. This is followed by nausea, cough, photophobia, and dizziness. The patient may be confused. There is often a rash. The incubation period before the onset of the first symptoms is about a week (though it can be shorter or longer). After the onset of disease, symptoms last a few days and then resolve.

After a week or two, the symptoms reoccur and in the absence of treatment, recurrence continues for several more episodes. As the fever resolves, the patient may go through a crisis in which first there is a high fever accompanied by confusion and delirium. This "chill phase" lasts up to half an hour, Then there is the "flush phase" in which the temperature drops accompanied by profuse sweating and sometimes a drop in blood pressure.

Diagnosis

Microscope smears of blood, bone marrow or cerebrospinal fluid stained with Giemsa or acridine orange. Serologic testing is also available.

Treatment

Antibiotics are used and symptoms resolve a few days. There can, however, be long-term sequelae including heart and kidney problems, peripheral nerve involvement, ophthalmia, and abortion. Without treatment mortality may be up to 10% of patients.

HARD TICKS

Dog tick (*Dermacentor variabilis*) which is found east of the Rocky Mountains and in some areas of the Pacific coast states

Rocky Mountain Wood Tick (*Dermacentor andersoni*). As its name suggests, it is found in the Rocky Mountains and also in southwest Canada

Deer Tick (Black-legged tick) (*Ixodes scapularis*). This occurs in the north east and north central United States.

Western black legged tick (*Ixodes pacificus*) which is found in the Pacific coast states of the United States.

Brown dog tick (*Rhipicephalus sanuguineus*). Also known as the red dog tick. This is found world-wide (all over the US and also southeast Canada) and can complete its entire life cycle indoors. It primarily infests dogs but can feed on other mammals including man.

Lone star tick (*Amblyomma americanum*) found in south eastern and south central United States.

HARD TICKS CARRIERS

ROCKY MOUNTAIN SPOTTED FEVER

There are several hundred reported cases of Rocky Mountain Spotted Fever each year in the United States (ranging, during the past half century, from a low of about 200 to more than 1200 in the early 1980's).

The numbers are again rising. Most at risk are children under 15 years of age. Usually, cases occur in the summer because of higher numbers of ticks and more frequent contact of humans with ticks.

Epidemiology

The causative agent, *Ricketsia rickettsii,* is carried by the Brown Dog tick and the Rocky Mountain Wood Tick (the two *Dermacentor* species in the United States). Contrary to its name, only a small proportion of cases are actually reported from the Rocky Mountain states. The highest number of cases in the United States occurs in the south-east and south central regions with the greatest incidence in Oklahoma and North Carolina.

Elsewhere in central and south America, *Rhipicephalus sanguineus* and *Amblyomma cajennense* carry *Ricketsia rickettsii.* The disease is known as fiebre manchada in Mexico; São Paulo fever or fiebre maculosa in Brazil; tick typhus or Tobia fever in Colombia.

Symptoms

R. rickettsii is a small bacterium that grows inside cells, particularly endothelial cells that form the walls of small blood vessels. The disease is characterized by nausea, appetite-loss, fever, myalgia and headache. These are followed, 3 to 5 days after the tick bite, by the characteristic rash which results from leakage of the blood vessels as a result of infected and dying endothelial cells.

Initially, the rash is formed of small, flat, pink, non-itchy macules (spots) on the wrists, forearms, and ankles. Subsequently, the macules become raised on the skin and there is pain (in the abdomen and joints) and diarrhea. The characteristic red rash, which occurs in up to 60% of patients, is found at the extremities (the palms and soles of feet). A minority of patients never progress to this stage.

Laboratory tests show thrombocytopenia, hyponatremia and/or elevated levels of liver enzymes. Severe cases require hospitalization and can result in paralysis of the extremities and may even be life-threatening. Very severe sequelae include gangrene that may result in amputation, deafness, and incontinence.

Treatment

Treatment is by antibiotics (doxycycline). Since, if untreated, Rocky Mountain Spotted Fever can be fatal, treatment should be started as soon as this disease is suspected and before any diagnosis is confirmed by laboratory tests.

Laboratory Detection

Serologic assays including indirect immunofluorescence microscopy.

Prevention

Clothes that cover body and anti-tick sprays (often containing DEET) are most often used. It is best to keep away from heavily tick-infested areas. If ticks are discovered on the body, they should be removed immediately using fingers or tweezers.

EHRLICHOSIS

Human Ehrlichosis

Human ehrlichosis is carried by *Dermacentor variabilis* and by *Amblyomma americanum* and is caused by a number of bacteria of the *Ehrlichia* family, in the United States principally by *Ehrlichia chaffeensis*. These bacteria are small gram-negative organisms that infect leukocytes. As with many tick-borne diseases, incidence follows vector distribution with higher incidence during the summer months when tick populations and contact with them are higher. The number of cases has been increasing.

Symptoms

After an incubation of period of a week to 10 days, the patient presents with myalgia, headache and general malaise. There can also be nausea, vomiting, diarrhea, cough, joint pains and the patient may be confused. Sometimes, there is a rash but this is normally only in pediatric cases. If left untreated, more severe manifestations of the infection can occur, including prolonged fever, renal failure, disseminated intravascular coagulopathy, meningoenceph-alitis, adult respiratory distress syndrome, seizures, or coma. Mortality rate at this stage is 2 - 3%. More at risk are immune-suppressed patients.

Diagnosis

Microscopy using blood smears or serology to detect anti-*Ehrlichia* antibodies can be used.

Human Granulocytic Ehrlichiosis

Human granulocytic ehrlichiosis is caused by a species of *Ehrlichia* similar to species found in animals (*Ehrlichia equi* and *Ehrlichia phagocytophila*) and is transmitted by blacklegged ticks (*Ixodes scapularis*) and western blacklegged ticks (*Ixodes pacificus*).

TULAREMIA

This is also carried by the two *Dermacentor* species. Tularemia is caused by the bacterium Francisella tularensis, which is carried by rodents, rabbits and hares; as a result tularemia is otherwise known as rabbit fever. One of the several ways that humans can be infected is by being bitten by a tick that has acquired the bacterium after biting one of these animals; however, it can also be inhaled during the handling of infected rodents. There have been no reports to person-to-person transmission. Francisella tularensis *is very infectious.* Tularemia occurs all across the continental United States but is relatively rare, with about 200 cases being reported each year.

Symptoms

The symptoms of tularemia, which can be fatal if untreated, vary according to the route by which the infection was acquired; often the patient experiences swollen lymph glands, skin ulcers inflammation of the eyes and throat, diarrhea. This may be followed by atypical pneumonia, pleuritis, and hilar lymphadenopathy. Inhaled tularemia results in rapid fever, chills, headache, myalgia, joint pain, dry cough, and progressive weakness. The pneumonia can result in respiratory distress and failure with blood in the sputum.

Diagnosis is initially from the symptoms but confirmatory laboratory tests using Gram or other stains or immunofluorescence microscopy are used to visualize the infecting bacteria.

Treatment

Oral antibitotc treatment using streptomycin, gentamycin,

tetracyclines (e.g. doxycycline) or fluoroquinolones, (e.g. ciprofloxacin) is the major form of therapy. Streptomycin or gentamicin can be used intravenously. There is a vaccine that is made from avirulent *F. tularensis* biovar palaearctica (type B). In addition, antibiotics can be used as post-exposure prophylaxis before the onset of symptoms if infection by *F. tularensis* is suspected.

Fever

Various farm animals (cattle sheep goats etc) are the primary carriers of the bacterium *Coxiella burnetii* which causes Q fever. Spread to humans is usually via inhalation of dust containing dried urine, feces etc of infected animals. However, less commonly, the bacterium can be transmitted via the bite of *Dermacentor* ticks. Ingestion of contaminated milk can also lead to infection. *C. burnetii* infects macrophages and survives in the phagolysosome, where the bacteria multiply. The bacteria are released by lysis of the cells and phagolysosomes.

Symptoms

Chronic Q fever

If the patient fails to resolve the infection, chronic Q fever results. This can occur a few months after primary infection but can also occur many years later. Endocarditis of the aortic heart valves is the major problem that arises. This usually occurs in people with heart valve disease but also at risk are transplant, cancer and kidney disease patients. The chronic form of Q fever has a fatality rate of about 60 - 70%.

Diagnosis

Serology to determine the presence of antibodies against *Coxiella burnetii* is used.

Treatment

Antibiotics such as doxycyline are used to treat acute Q fever. For chronic Q fever, two protocols have been investigated: doxycycline along with quinolones for at least 4 years and doxycycline with hydroxychloroquine for 1.5 to 3 years. There is a vaccine used in Australia for persons who may come in contact with C. *burnettii* but it is not commercially available in the United States.

Acute Q fever

Many patients, about half, show no signs of infection but in others after an incubation period of 1 - 2 weeks, there is a sudden onset of fever, headache, general malaise, myalgia, sore throat, chills, sweats, non-productive cough, nausea, vomiting, diarrhea, abdominal pain, and chest pain. The patient may also appear confused. Many patients go on to the symptoms of pneumonia and hepatitis but most recover in a month or two without treatment although acute Q fever has a mortality rate of 1-2%.

LYME DISEASE

Lyme disease is caused by the spirochete bacterium, *Borrelia burgdorferi,* which typically infects small mammals in the northeast and north central United States. It is transmitted to humans by *Ixodid* black legged ticks (deer ticks). There are over 20,000 cases per year in the United States making it the most common tick-borne disease in North America. The disease was first described from the town of Old Lyme in Connecticut but is found on both the east and west coasts and in the Mississippi valley. In Europe, a similar disease is caused by *Borrelia garinii* or *Borrelia afzelii.*

Symptoms

Fever, headache and malaise and a characteristic rash named erythemia migrans which can occur in a few days but sometimes only after a few weeks, are typical of Lyme Disease. The rash (which is usually not painful) often has a bull's eye appearance since as it grows (up to 30 cm across) the central region clears. If left untreated, the infection spreads and can result in Bell's Palsy (partial paralysis of muscles in one or both sides of the face), meningitis, heart palpitations and severe joint pain.

These symptoms usually resolve in a few weeks but after several months about 60% of patients will get severe joint swelling and arthritis. A small minority may also get neurologic symptoms (tingling of the extremities, shooting pains, numbness)

Treatment

Early administration with antibiotics (doxycycline, amoxicillin, or cefuroxime axetil) is recommended. Some patients continue with neurological and muscle pain problems even after antibiotic treatment. It is not known what causes these but they may be autoimmune in nature.

SOUTHERN TICK-ASSOCIATED RASH ILLNESS

This rash is similar to that seen in Lyme disease. The causative organism is not known but it is not *Borrelia burgdorferi,* the Lyme disease agent. The lone star tick, *Amblyomma americanum,* is the transmission vector.

Symptoms

Malaise, fever, myalgia, arthralgia and a "bulls eye" rash at the site of the tick bite. There are no chronic neurological symptoms as are seen with Lyme disease.

Treatment

The usual oral antibiotics are used and the symptoms quickly resolve.

BABESIOSIS

Babesiosis is carried by species of *Ixodes* including the deer tick (*Ixodes scapularis*) in the north and mid-west of the United States and in other countries, including Europe. *Babesia microti* is the usual causative organism and is a hemoprotozoan (i.e. it circulates in the bloodstream). Normally, the two hosts of *Babesia microti* are ticks and peromyscus mice (*Peromyscus leucopus*). The tick infects the mice with sporozoites, which reproduce asexually in erythrocytes.

These escape to the blood stream where they may form

male and female gametes that are taken up by the tick during a blood meal. In the tick, the gametes fuse and go through a sporogonic cycle to form more sporozoites. Humans can also acquire sporozoites when bitten by an infected tick and are usually dead-end hosts but babesiosis has been transmitted to other humans via blood transfusions.

In most cases, infection is asymptomatic but after a week to a month, symptoms can appear. These include fever, chills, sweating, myalgias and fatigue. In severe cases, hepatosplenomegaly and hemolytic anemia can occur. Normally, the patient recovers, although severe cases occur in immuno-compromised patients and the elderly.

Disease cause by another protozoan, *Babesia divergens,* can cause more severe and sometimes fatal cases of babesiosis.

Diagnosis

Diagnosis is by serology, immunofluorescence microscopy and by direct observation of the parasite in blood smears in which "Maltese Cross"-like inclusions in erythrocytes are seen. These consist of four budding merozoites attached together.

Treatment

Usual antibiotics used are clindamycin plus quinine or atovaquone plus azithromycin.

CRIMEAN-CONGO HEMORRHAGIC FEVER

This is caused by a *Nairovirus,* a member of the *Bunyaviridae.* It is found in Eastern Europe and throughout the Mediterranean areas of southern Europe, the Middle East, Africa, northwestern China, central and south Asia. *Ixorid* ticks (genus *Hyalomma*) spread the virus, which is also carried by numerous species of domestic and wild animals. Person to person transmission through infected blood and other body fluids has been documented.

Symptoms

Initially, the patient presents with headache, high fever, back pain, joint pain, stomach pain, and vomiting. There may be flushing, red eyes and throat and small red spots called petechiae on the palate. Hemorrhage ensues after a few days and lasts for a few weeks. This is indicated by severe bruising, nosebleeds, and failure to stop bleedings after a cut or injection. Slow recovery often ensues but mortality can be as high as 50%.

Treatment

Since this is a viral disease, treatment is largely supportive with particular attention to electrolyte balance. Ribavirin has been used. An inactivated vaccine has been used in Eastern Europe.

COLORADO TICK FEVER

This is sometimes confused with a mild case of Rocky Mountain Spotted Fever but Colorado Tick Fever is caused by a *coltivirus*, a member of the reoviruses. They are endemic to north western North America and are found in *Ixodid* ticks. The virus distribution closely matches that of its vector, *Dermacentor andersoni.*

Person-to-person transmission can occur by blood. Prolonged viremia observed in humans and rodents is due to the intraerythrocytic location of virions, which protects them from immune clearance.

Symptoms

Infection results in abrupt fever, chills, headache, retro-orbital pain, photophobia, myalgia, abdominal pain, and maıaise. Sometimes fever can be diphasic or triphasic, usually lasting for 5 to 10 days. Severe forms of the disease that involve infection of the central nervous system or hemorrhagic fever, pericarditis, myocarditis, and orchitis have been rarely observed, mainly in children. Severity is sufficient to result in hospitalization of approximately 20% of patients. There has been evidence of transmission from mother to child.

FAR EASTERN TICK-BORNE ENCEPHALITIS

This is caused by a flavivirus which is spread by *ixodid*

ticks (*Ixodes persulcatus, I. ricinus* and *I. cookie*). Small animals are the reservoir and the virus is endemic to the former Soviet Union and parts of eastern and central Europe.

Symptoms

About two weeks after infection, there is a mild influenza-like disease that normally resolves in a few days but which can be followed by meningitis and meningoencephalitis in about one third of cases. In some cases, there may be partial paralysis and mortality may be as high as 25%.

Treatment

Supportive care is normal. There is a vaccine of killed virus that is available in Europe.

TICK-BORNE ENCEPHALITIS

This disease results from infection by tick-borne encephalitis virus, which is a member of the *Flaviviridae.*

Symptoms

Tick-borne encephalitis starts as mild influenza-like symptoms with fever accompanied by leuko- and thrombocytopenia. This resolves within a few days. However, about one third of patients develop meningitis and meningoencephalitis. This can, in a few cases, be followed by paralysis. The European form of the disease has a mortality rate of under 5%. Most patients recover but about a third may have long-lasting neurological problems.

Treatment

Supportive is indicated. There is an experimental killed vaccine in Europe. In Sweden TBE vaccination is recommended for residents of and regular visitors to TBE endemic areas.

KYASANUR FOREST DISEASE

This disease is similar to Russian spring-summer encephalitis and is also caused by flaviviruses. It is found only in the Kyasanur forest of Northern India. The disease occurs during the dry season as its tick vector (*Haemaphyalis spinigera*) begins to feed on humans. Local carriers are shrews and monkeys.

LOUPING ILL VIRUS

This is found in the British Isles and is caused by a flavivirus that is carried by pheasants and sheep, among other animals. It can infect many hosts via the tick vector, *Ixodes ricinus.* It causes mild encephalitis that gives the infected animal an unusual gait (hence its name). However, it can kill livestock and humans not given proper supportive care.

DISEASE CAUSED DIRECTLY BY HARD TICKS

Tick Paralysis

In addition to being carriers of disease-causing microorganisms, some ticks (*Amblyomma americanum* and the two *Dermacentor* species) can cause tick paralysis. This is a rare disease caused by toxin in the saliva of the tick and results in an acute, ascending, flaccid paralysis caused by reduced acetyl choline or motor neuron action potentials.

The paralysis, which is not associated with pain, starts a few days after the bite and comes on gradually over a period of days. The paralysis resolves surprisingly rapidly, usually within a day of the removal of the tick but if the tick is not removed the mortality rate, as a result of respiratory paralysis, can be as high as 10%. Tick paralysis can be confused with other acute neurologic disorders or diseases (e.g., Guillain-Barré syndrome or botulism).

Chapter 8

Yeasts

Yeasts are single-celled budding organisms. They do not produce mycelia. The colonies are usually visible on the plates in 24-48 hours. Their soft, moist colonies resemble bacterial cultures rather than molds. There are many species of yeasts which can be pathogenic for humans. We will discuss only the two most significant species: *Candida albicans* and *Cryptococcus neoformans*.

CANDIDIASIS

There are many species of the genus Candida that cause disease. The infections caused by all species of Candida are called candidiasis. *Candida albicans* (2 and 3) is an endogenous organism. It can be found in 40-80% of normal human beings. It is present in the mouth, gut, and vagina. It may be present as a commensal or a pathogenic organism. Infections with Candida usually occur when a patient has some alteration in cellular immunity, normal flora or normal physiology. Patients with decreased cellular immunity have decreased resistance to fungal infections.

Prolonged antibiotic or steroid therapy destroys the balance of normal flora in the intestine allowing the endogenous Candida to overcome the host. Invasive procedures, such as cardiac surgery and indwelling catheters, produce alterations in host physiology and some of these patients develop Candida infections. Although it most frequently infects the skin and mucosae, Candida can cause pneumonia, septicemia or endocarditis in the immuno-compromised patient. The establishment of infection with

Candida species appears to be a property of the host - not the organism. The more debilitated the host, the more invasive the disease.

The clinical material to be sent to the lab depends on the presentation of the disease: blood cultures, vaginal discharge, urine, feces, nail clippings or material from cutaneous or mucocutaneous lesions. Candida is a polymorphic yeast, i.e., yeast cells, hyphae and pseudohyphae are produced. It has been shown that Candida needs a transcription repressor to maintain the yeast form.

This ability to assume various forms may be related to the pathogenicity of this organism. The yeast form is 10-12 microns in diameter, gram positive, and it grows overnight on most bacterial and fungal media. It also produces germ tubes and pseudohyphae may be formed from budding yeast cells that remain attached to each other.

Spores may be formed on the pseudomycelium. These are called chlamydospores and they can be used to identify different species of Candida. Some mycologists think that the pseudomycelial form represents a more invasive form of the organism. The species are identified by biochemical reactions. The organism occurs world-wide. The drugs of choice for systemic infection are itraconazole and fluconazole. If an artificial heart valve or in-dwelling catheter becomes infected, it must be replaced. Drug therapy alone will not suppress the organism if the foreign body remains in the host. This resistance is due to biofilms which we will discuss later

CRYPTOCOCCOSIS

Cryptococcosis manifests itself most commonly as meningitis but in recent years many cases of pulmonary disease have been recognized. *C. neoformans* is a very distinctive yeast. The cells which are spherical and 3-7 microns in diameter, produce buds which characteristically are narrow-based and the organism is surrounded by a polysaccharide capsule. There is evidence that the capsule may suppress T-cell function and can be considered a virulence factor. *C. neoformans* also produces an enzyme called phenoloxidase which appears to be another virulence factor. The ecological niche of *C. neoformans* is pigeon and chicken droppings. However, although this organism can be easily recovered from pigeon droppings, a direct epidemiological link has yet to be established between exposure to pigeon droppings and a specific human infection. Infection and disease production is probably a property of the host–not the organism. The source of human infection is not clear. This organism is ubiquitous, especially in areas like abandoned buildings contaminated with pigeon droppings.

The portal of entry is the respiratory system. Evidence is developing which indicates that the initial exposure may be many years prior to the manifestation of disease. The organism can be sequestered for this time. Infection may be subacute or chronic. The highly fatal meningoencephalitis caused by *C. neoformans* has a prolonged evolution of several months. The patients symptoms may begin with vision problems and headache, which then progress to delirium, nuchal rigidity leading to coma and death unless the physician is thinking about cryptococcus and does a spinal tap for diagnosis and institutes aggressive therapy.

The CSF is examined for its characteristic chemistry (elevated protein and decreased glucose), cells (usually monocytes), and evidence of the organism. The latter is measured by the visual demonstration of the organism (India Ink preparation) or by a serologic assay for the antigen of *C. neoformans*. The India Ink test, which demonstrates the capsule of this yeast, is supplemented by the latex agglutination test for antigen which is more sensitive and more specific. The Latex Agglutination test measures antigen, NOT antibody.

A decreasing titer indicates a good prognosis, while an increasing titer has a poor prognosis. When you consider Cryptococcosis, think of Capsules and CNS disease. In addition to causing meningitis, *C. neoformans* may also infect lungs and skin. The disease in the lungs and skin is characterized by the formation of a granulomatous reaction with giant cells. As with other fungal diseases, there has been an increase in the recognition of pulmonary infection.

The yeast may also form a mass in the mediastinum called a cryptococcoma. The geographical distribution of this organism is world-wide. The clinical material sent to the lab is CSF, biopsy material, and urine (for some unexplained reason the organism can be isolated from the urine in both the CNS and systemic infections). This organism will grow overnight on bacterial or fungal media at 37 C. but growth is a little slower at room temperature. In culture the organism grows as creamy, white, mucoid (because of the capsule) colonies.

Growth in culture is usually visible in 24 to 48 hours. As the culture ages, it turns brown due to a melanin produced by the phenoloxidase. The organism is a round, single cell, yeast surrounded by a capsule. Identification is based on physiological reactions. Pathologists use a mucicarmine stain, which stains the capsule, to identify the organism in tissue sections.

There is usually little or no inflammatory response. The Direct Fluorescent Antibody test identifies the organism in culture or tissue section specifically, by causing the yeast cell wall to stain green. To test the patient's serum there are 3 serologic tests: The Indirect Fluorescent Antibody test, the Tube Agglutination test for antibody, and the Latex Agglutination test for antigen.

The latex agglutination test can be used as a prognostic test. As the patient improves, the serum antigen titer will also decrease. The drugs of choice to treat cryptococcus infection are amphotericin B and 5-Fluorocytosine (5-FC). 5-FC is an oral drug. If it is given as the only treatment, there are relapses so most physicians use both drugs simultaneously. Actually, these two drugs are synergistic, and thus, their association is advantageous.

SUPERFICIAL MYCOSES

The superficial (cutaneous) mycoses are usually confined to the outer layers of skin, hair, and nails, and do not invade living tissues. The fungi are called dermatophytes. Dermatophytes, or more properly, keratinophilic fungi, produce extracellular enzymes (keratinases) which are capable of hydrolyzing keratin.

CLINICAL MANIFESTATIONS

Tinea means "ringworm" or "moth-like". Dermatologists use the term to refer to a variety of lesions of the skin or scalp.

- *Tinea corporis*: Small lesions occurring anywhere on the body.
- *Tinea pedis:* "Athlete's foot". Infection of toe webs and soles of feet.
- *Tinea unguium (onychomycosis):* Nails. Clipped and used for culture.
- *Tinea capitis:* Head. Frequently found in children
- *Tinea cruris:* "Jock itch". Infection of the groin, perineum or perianal area.
- *Tinea barbae*: Ringworm of the bearded areas of the face and neck.
- *Tinea versicolor:* Characterized by a blotchy discoloration of skin which may itch.

Up to 25% of the general population may have this lesion at any one time. Diagnosis is usually possible by direct microscopic examination of KOH-treated skin scrapings which show a typical aspect of mycelia and spores described as "spaghetti and meatballs." Caused by *Malassezia furfur*.

ECOLOGY

The dermatophytes (skin plants) causing human infections may have different natural sources and modes of transmission.

- *Anthropophilic:* Usually associated with humans only; transmission from man to man by close contact or through contaminated objects.
- *Zoophilic:* Usually associated with animals; transmission to man by close contact with animals (cats, dogs, cows) or with contaminated products.
- *Geophilic:* Usually found in the soil, transmitted to man by direct exposure. Knowledge of the species of dermatophyte and source of infection are important for proper treatment of the patient and

control of the source. Invasion by zoophilic or geophilic organisms may cause inflammatory disease in man.

Geographic distribution: Dermatophytes occur worldwide, but some species have geographically limited distribution.

ETIOLOGIC AGENTS

There are three genera of dermatophytes:

Epidermophyton floccosum

These infect skin and nails and rarely hair. They form yellow-colored, cottony cultures and are usually readily identified by the thick, bifurcated hyphae with multiple smooth, club-shaped macroconidia.

Trichophyton species

These infect skin, hair and nails. Rarely can cause subcutaneous infections, in immunocompromised individuals. Take 2-3 weeks to grow in culture. The conidia are large (macroconidia), smooth, thin-wall, septate and pencil-shaped; colonies a re a loose aerial mycelium which grow in a variety of colors. Identification requires special biochemical and morphological techniques. Trichophyton rubrum is presently the most common cause of tinea in South Carolina.

Microsporum species

These may infect skin and hair, rarely nails. Its prevalence has decreased significantly. When prevalent (15-20 years ago), this organism could be easily identified on the scalp because infected hairs fluoresce a bright green colour when illuminated with a UV-emitting Wood's light. The loose, cottony mycelia produce macroconidia which are thick-walled, spindle-shaped, multicellular, and echinulate (spiny). Microsporum canis is one of the most common dermatophyte species infecting humans.

THERAPY

Skin infections can be treated (more or less successfully) with a variety of drugs, such as:

Tolfnatate (Tinactin) available over the counter: Topical

Clotrimazole: Topical

Miconazole: Topical.

Ketoconazole seems to be most effective for tinea versicolor and other dermatophytes.

Itraconazole: Oral

Terbinifine (Lamisil): Oral, topical.

For skin and Nail infections.

Morpholines: Oral

For infections involving the scalp and particularly the nails, griseofulvin is commonly used. This antimycotic must be incorporated into the newly produced keratin layer to form a barrier against further invasion by the fungus. This is a very slow process requiring oral administration of the drug for long periods - up to 6 to 9 months for fingernail infections and 12 to18 months for toenail infections. Itraconazole and terbinafine are the drugs of choice for onychomycoses.

THE IDENTIFICATION REACTION

Patients infected with a dermatophyte may show a lesion, often on the hands, from which no fungi can be recovered or demonstrated. It is believed that these lesions, which often occur on the dominant hand (i.e. right-handed or left-handed), are secondary to immunological sensitization to a primary (and often unnoticed) infection located somewhere else (e.g. feet). These secondary lesions will not respond to topical treatment but will resolve if the primary infection is successfully treated.

FILAMENTOUS FUNGI

CHROMOBLASTOMYCOSIS

A chronic, localized infection infection of subcutaneous tissues caused by several species of dematiaceous fungi. The 3 most common agents are:

- Fonsecaea pedrosoi
- Cladosporium carrionii and
- Phialophora verrucosa

These fungi, recognized by a variety of names, are saprobes located in soil and decaying vegetation. The route of entry is usually by trauma. The lesions are sub-cutaneous and the surface can be flat or verrucous. The lesions take several years to develop. These organisms are called dematiaceous fungi, because they have a black colour in the mycelium cell wall (in culture and in tissue). In tissue these fungi form sclerotic bodies which are the reproductive forms dividing by fission. These organisms induce a granulomatous reaction.

The etiologic agents of chromoblastomycosis are septate, mold-like, branching, darkly pigmented which produce asexual fruits called conidia. We identify these fungi in culture by the shape and formation of the conidia. The fungi have a world-wide distribution especially in warmer climates like the tropics or the southern U.S.

The melanin in the pigment may be a virulence factor. These organisms are distributed world-wide. There is no really successful therapy. Excision and local heat have been used with some success. Flucytosine (5-FC), thiabendazole and itraconazole have also been used to treat (or control) this disease. There are no serological tests to aid in the diagnosis.

MYCETOMA

Mycetomas (fungous tumors) are also chronic, subcutaneous infections. These are called eumycotic mycetoma (tumors caused by the TRUE fungi as opposed to those caused by actinomycetes). These tumors frequently invade contiguous tissue, particularly the bone. A diagnosis of the etiologic agent is essential for patient management because the prognosis and therapy differs. Mycetoma characteristics:

- *Tumefaction*: Swelling
- *Granules*: A variety of colors (white, brown, yellow, black)
- Draining sinus tracts

The three most common etiologic agents are:

- Madurella mycetomatis
- Exophiala jeanselmei
- Pseudallescheria boydii

Clinical specimens for diagnosis:

- *Pus*: With granules.
- *Tissue*: For histological examination.

The colour, size and texture of the granules are an aid in the diagnosis of mycetomas. The agents of mycetoma are all filamentous fungi which require 7-10 days for visible growth on the culture media and then another several days for specific identification. These fungi are identified by the colonial morphology, conidia formation and biochemical reactions. The species of fungi cannot be distinguished in histopathological tissue sections. Treatment is very difficult, but ketoconazole and itraconazole have been used with some success.

ZYGOMYCOSIS

Also known as mucormycosis and phycomycosis. Zygomycosis is an acute inflammation of soft tissue, usually with fungal invasion of the blood vessels. This rapidly fatal disease is caused by several different species in this class. The zygomycetes, like the Candida species, are ubiquitous and rarely cause disease in an immunocompetent host. Some characteristic underlying conditions which cause susceptibility are: diabetes, severe burns, immunosuppression or intravenous drug use.

The three most common genera causing this clinical entity are:

- Rhizopus species
- Mucor species
- Absidia species

Characteristics: world-wide distribution, commonly in soil, food, organic debris, seen on decaying vegetables in the refrigerator and on moldy bread. Rhinocerebral infections are common. This disease is frequently seen in the uncontrolled diabetic. Typical case: An uncontrolled diabetic patient comes to ER (may be comatose depending on the state of diabetes) and a cotton-like growth is observed on the roof of the mouth or in the nose. These are the hyphae of the organism. If untreated, the patient will die within a few hours or days. What do you do to help this patient first? Controlling the diabetic state is most important before administering amphotericin. These fungi have a tendency to invade blood vessels (particularly arteries) and enter the brain via the blood vessels and by direct extension through the cribiform plate. This is why they cause death so quickly.

Culture: A rapid growing, loose, white mold which is visible in 24 to 48 hours. With age, and the formation of sporangia, the colony becomes dark gray. The sporangia contain the dark spores. The mycelium is, wide (10-15 microns), ribbon-like and non-septate (coenocytic). This same appearance is clear in tissue sections. The species are identified by the morphology in culture.

There is an immunodiffusion test available, but the physician cannot wait for these results before instituting rapid, vigorous intervention. The diagnosis and treatment must be immediate and based primarily on clinical observations.

ASPERGILLOSIS

Aspergilli produce a wide variety of diseases. Like the zygomycetes, they are ubiquitous in nature and play a significant role in the degradation of plant material as in composting. Similar to Candida and the Zygomycetes, they rarely infect a normal host. The organism is distributed worldwide and is commonly found in soil, food, paint, air vents. They can even grow in disinfectant. There are more than one hundred species of aspergilli The most common etiologic agents of aspergillosis in the United States:

- *Aspergillus fumigatus*
- *Aspergillus niger*
- *Aspergillus flavus*

There are three clinical types of pulmonary aspergillosis:

- *Allergic*: Hypersensitivity to the organism. Symptoms may vary from mild respiratory distress to alveolar fibrosis.
- *Aggressive tissue invasion*: Primarily a pulmonary disease, but the aspergilli may disseminate to any organ. They may cause endocarditis, osteomyelitis, otomycosis and cutaneous lesions.

- Fungus ball which is characteristically seen in the old cavities of TB patients. This is easily recognized by x-ray, because the lesion (actually a colony of mold growing in the cavity) is shaped like a half-moon (semi-lunar growth). The patients may cough up the fungus elements because the organism frequently invades the bronchus. Chains of conidia can sometimes be seen in the sputum.
- *Culture:* Aspergilli require 1-3 weeks for growth. the colony begins as a dense white mycelium which later assumes a variety of colors, according to species, based on the colour of the conidia. The hyphae are branching and septate. Species differentiation is based on the formation of spores as well as their colour, shape and texture.
- *Histopathology:* The septate hyphae are wide and form dichotomous branching, i.e., a single hypha branches into two even hyphae, and then the mycelium continues branching in this fashion.
- *Serology:* There is an excellent serological test for aspergillosis which is an Immunodiffusion test. There may be 1 to 5 precipitin bands. Three or more bands usually indicate increasingly severity of the disease. i.e., tissue invasion.
- *Treatment:* Itraconazole and Amphotericin B.

DIMORPHIC FUNGI

BLASTOMYCOSIS (BLASTOMYCES DERMATITIDIS)

Most of the systemic fungi have a specific niche in nature where they are commonly found. *Blastomyces dermatitidis* survives in soil that contains organic debris (rotting wood, animal droppings, plant material) and infects people collecting firewood, tearing down old buildings or engaged in other outdoor activities which disrupt the soil. In addition to an ecological niche, most fungi that cause systemic infections have a limited geographic distribution where they occur most frequently. Blastomycosis occurs in eastern North America and Africa. The vast majority of patients with blastomycosis in South Carolina are infected in the northern part of the state, above the Fall Line (Augusta, GA, Aiken, Columbia, Cheraw, Raleigh, NC).

Blastomycosis is a chronic granulomatous disease which means that it progresses slowly. Although the pulmonary and skin involvement is the most common, *B. dermatitidis* frequently affects bone, prostate and other organs. More frequently blastomycosis presents as a cutaneous or a respiratory disease. The cutaneous lesions may be primary (usually self-limiting) or secondary (a manifestation of systemic disease).

The patient who presents with a complaint of respiratory symptoms will frequently remark about loss of appetite, loss of weight, fever, productive cough, and night sweats. While these symptoms resemble those of TB, it is not this disease. The X-ray shows obvious pulmonary disease.

To make the specific diagnosis, the physician must be aware of blastomycosis. Sputum sent to the lab for "culture" will not grow the organism. The lab must be alerted to look for fungal organisms or to look specifically for blastomyces. Some patients have a sub-clinical or "flu-like" response to infection. *B. dermatitidis* can frequently be demonstrated in a KOH preparation of pus from a skin lesion.

A typical cutaneous lesion shows central healing with microabscesses at the periphery. A pus specimen may be obtained by nicking the top of a microabscess with a scalpel, obtaining the purulent material and making the diagnosis in 5 min. by microscopic examination with KOH. This organism has a characteristic appearance of a double contoured wall with a single bud on a wide base (s 2 - 5). There are no specific virulence factors for *B. dermatitidis*. Laboratory specimens depend on the manifestation of the disease: If there are skin lesions, send skin scrapings or pus.

If there is pulmonary involvement, send sputum. Other specimens include biopsy material and urine. Occasionally, the organism can be isolated from urine as it often infects the prostate.

MYCOLOGY

If you request a fungus culture from the microbiology lab, they will incubate the cultures at 37 degrees C and at 25 degrees C because most of the significant pathogenic fungi are dimorphic.

A culture of *B. dermatitidis* takes 2 to 3 weeks to grow at 25 degrees C. It appears as a white, cottony mold (mycelium) on Sabouraud dextrose agar. Most specimens for fungus culture are plated on Sabouraud's dextrose agar. Microscopically, the mycelia and the fruiting bodies are evident. However, the mold cannot be identified by its fruiting bodies. The fruiting bodies are called microconidia, but they are not distinctive. Other fungal saprophytes and pathogens have similar conidia. At 37 degrees C the yeast form grows in about 7-10 days. It appears as a buttery-like, soft colony with a tan colour.

Microscopically, we see the typical yeast form of a thick wall and a single bud with a WIDE BASE. This wide base is characteristic of *B. dermatitidis*, and it is important to be able to recognize this. The cells are 12-15 microns in diameter. The yeast will convert to the mycelial form when incubated at 25 degrees C, taking from 3 to 4 days up to a few weeks. Similarly, the mycelial growth can be converted to yeast form when incubated at 37 degrees C.In the past, the only way to identify the dimorphic fungi was to convert from one form to the other, but now it is possible to take the mycelial growth (which is the easiest to grow), and confirm the isolate with a DNA probe in a matter of hours.

HISTOPATHOLOGY

B. dermatitidis produces both a granulomatous and suppurative tissue reaction

SEROLOGY

There are three serological tests used for blastomycosis:

Immunodiffusion test

This requires 2 to 3 weeks to become positive. This test is

positive in about 80% of the patients with blastomycosis. When it is positive, there is close to 100% specificity.

Complement Fixation test

This test requires 2 to 3 months after the onset of disease to develop detectable antibody. Besides the long delay before there is measurable antibody, another disadvantage of the C-F is that it cross reacts with other fungal infections (coccidioidomycosis and histoplasmosis). The advantage is that it is a quantitative test. The physician can follow the patient's response to the disease by monitoring the antibody titer.

Enzyme Immunoassay

The latter test has met with mixed acceptance by mycologists. However it is easy to perform and antibody is detected early in the disease process. Amphotericin B remains is the drug of choice (DOC) although it is very toxic and must be administered intravenously for several weeks. Ketoconazole is also being used in mild cases.

HISTOPLASMOSIS (HISTOPLASMA CAPSULATUM)

Histoplasmosis is a systemic disease, mostly of the reticuloendothelial system, manifesting itself in the bone marrow, lungs liver, and the spleen. In fact, hepatosplenomegaly is the primary sign in children, while in adults, histoplasmosis more commonly appears as pulmonary disease. This is one of the most common fungal infections, occurring frequently in South Carolina, particularly the northwestern portion of the state. The ecological niche of *H. capsulatum* is in blackbird roosts, chicken houses and bat guano.

Typically, a patient will have spread chicken manure around his garden and 3 weeks later will develop pulmonary infection. There have been several outbreaks in South Carolina where workers have cleared canebrakes which served as blackbird roosts with bulldozers. All who were exposed, workers and bystanders, contracted histoplasmosis.

Histoplasmosis is a significant occupational disease in bat caves in Mexico when workers harvest the guano for fertilizer. In the endemic area the majority of patients who develop histoplasmosis (95%) are asymptomatic. The diagnosis is made from their history, serologic testing or skin test. In the patients who are clinically ill, histoplasmosis generally occurs in one of three forms: acute pulmonary, chronic pulmonary or disseminated.

There is generally complete recovery from the acute pulmonary form (another "flu-like" illness). However, if untreated, the disseminated form of disease is usually fatal. Patients will first notice shortness of breath and a cough which becomes productive.

The sputum may be purulent or bloody. Patients will become anorexic and lose weight. They have night sweats. This again sounds like tuberculosis, and the lung x- ray also looks like tuberculosis, but today radiologists can distinguish between these diseases on the chest film (histoplasmosis usually appears as bilateral interstitial infiltrates). Histoplasmosis is prevalent primarily in the eastern U.S.

In S.C., a histoplasmin skin test survey of lifetime, one county residents, white males, 17 to 21 years old, was performed on Navy recruits.

The greatest number of positive skin tests appeared in the northwestern part of the state. A similar study of medical students conducted at Medical University of South Carolina, about 25 years ago, bore the same distribution.The skin test is not used for diagnostic purposes, because it interferes with serological tests. Skin tests are used for epidemiological surveys.

Clinical specimens sent to the lab depend on the presentation of the disease: Sputum or Bronchial alveolar lavage, if it is pulmonary disease, or Biopsy material from the diseased organ. Bone marrow is an excellent source of the fungus, which tends to grow in the reticulo-endothelial system. Peripheral blood is also a source of visualizing the organism histologically. The yeast (s 7-11) is usually found in monocytes or in PMN's.

Many times an astute medical technologist performing a white blood cell count will be the first one to make the diagnosis of histoplasmosis. In peripheral blood, *H. capsulatum* appears as a small yeast about 5-6 microns in diameter. (Blastomyces is 12 to 15 microns). Gastric washings are also a source of *H. capsulatum* as people with pulmonary disease produce sputum and frequently swallow their sputum.

MYCOLOGY

When it is grown on Sabouraud dextrose agar at 25 degrees C, it appears as a white, cottony mycelium after 2 to 3 weeks. As the colony ages, it becomes tan. In the mold form, Histoplasma has a very distinct spore called a tuberculate macroconidium. The tubercles are diagnostic, however there are some non-pathogens which appear similar. A medical mycologist will be able to distinguish them.

Grown at 37 degrees, C the yeast form appears. It is a white to tan colony. The yeast cell is 5-6 microns in diameter and slightly oval in shape. This is not diagnostic. To confirm the diagnosis, one must convert the organism from yeast to mycelium or vice-versa or use the DNA probe.

Serology for histoplasmosis is a little more complicated than for other mycoses, but it provides more information than blastomycosis serology.

There are 4 tests:

- Latex agglutination
- Complement Fixation
- Immunodiffusion
- EIA

Each of these serological tests has different characteristics that make them useful. The latex agglutination test is a very simple test involving agglutination in a test tube. The Ab is fairly specific and rises early in the disease (in the first 2 weeks), and disappears in about 3 months. The complement fixation test is like the one for blastomycosis, except there are 2 antigens, one to the yeast form of the organism and the other to the mycelial form. Some patients react to one form and not the other, while some individuals react to both.

The reason for the different responses is not clear. One disadvantage is that complement fixing antibody develops late in the disease, about 2 to 3 months after onset. A second disadvantage is that it cross reacts with other mycotic infections. An advantage of the C-F test is that it is quantitative, so the physician can follow the course of the disease by observing the titer of several samples. The interpretation of the immunodiffusion test is a little more complicated than with blastomycosis because there are two bands which may appear.

An H band indicates active disease and will appear in 2 to 3 weeks. An M band can indicate past or present disease, or result from a skin test. This is one reason why skin tests are not used for diagnosis because they can interfere with other tests. Skin tests will also affect the complement fixation test.

Recently, a radioimmunoassay for histoplasma polysaccharide antigen has been developed. This is a proprietary test so the evaluation of the results have been questioned. The drug of choice (DOC) is amphotericin B, with all its side effects. Itraconazole is now also being used.

COCCIDIOIDOMYCOSIS

Coccidioidomycosis is primarily a pulmonary disease. About 60% of the infections in the endemic area are asymptomatic. About 25% suffer a "flu-like" illness and recover without therapy. This disease exhibits the typical symptoms of a pulmonary fungal disease: anorexia, weight loss, cough, hemoptysis, and resembles TB. CNS infection with *C. immitis* is more common while it is less frequent with the other fungal diseases.

The ecological niche of *C. immitis* is the Sonoran desert, which includes the deserts of the Southwest (California, Arizona, New Mexico and Texas) and northern Mexico). It is also found in small foci in Central and South America.

Desert soil, pottery, archaeological middens, cotton, and rodent burrows all harbor *C. immitis*. *C. immitis* is a dimorphic fungus with 2 life cycles. The organism follows the SAPROPHYTIC cycle in the soil and the PARASITIC cycle in man or animals. The saprophytic cycle starts in the soil with spores (arthroconidia) that develop into mycelium. The mycelium then matures and forms alternating spores within itself. The arthroconidia are then released, and germinate back into mycelia. The parasitic cycle involves the inhalation of the arthroconidia by animals which then form spherules filled with endospores.

The ambient temperature and availability of oxygen appear to govern the pathway The organism can be carried by the wind and therefore spread hundreds of miles in storms so the distribution is quite wide. In 1978, cases were seen in Sacramento 500 miles north of the endemic area, from a dust storm in Southern California.

The spores of the organism are readily airborne. The cases that occur in South Carolina are usually in patients who have visited an endemic area and brought back pottery, or blankets purchase from a dusty roadside stand, or in Navy and Air Force personnel who were exposed when they were stationed in the endemic area.

The disease manifests itself after they are transferred to a base in South Carolina. A few interesting cases occurred in cotton mills in Burlington and Charlotte, N.C. The cotton, grown in the desert of the Southwest, was contaminated with the fungus and the mill workers inhaled the spores while handling the raw cotton and developed coccidioidomycosis.

CLINICAL SPECIMENS

Clinical specimens include sputum, pus from skin lesions, gastric washings, CSF, and biopsy material from skin lesions. *C. immitis* is a dimorphic fungus (19-23). Cultured on Sabouraud's agar at 25 degrees C it grows as a mold in 2 to 3 weeks. Characteristically, the mycelia develop arthroconidia. ("By their fruits ye shall know them"). It is a barrel-shaped (smaller at the edges, wider at the middle) asexual spore. Typically, the arthroconidia alternate with non spore-forming cells in the mycelium.

When grown *in vitro* at 37 degrees C, there is no yeast form!! *C. immitis* is a dimorphic fungus; *in vivo*, (pus or tissue) one sees the pathogenic or invasive form B which is a spherule.

The organism develops into spherules (30-60 microns) that are filled with endospores which are 3 to 5 microns in diameter. A spherule will develop endospores within, then break apart, releasing the endospores. This is the tissue form seen in pus or histological sections: spherules and loose endospores. They can also be seen in a KOH preparation of sputum. It is pathognomonic for coccidioidomycosis.

Histopathology

The inflammatory reaction is both purulent and granulomatous. Recently released endospores incite a polymorphonuclear response. As the endospores mature into spherules, the acute reaction is replaced by lymphocytes, plasma cells, epithelioid cells and giant cells.

Serology

There are four tests for diagnosis:

- Complement-Fixation
- Slide agglutination
- Immunodiffusion
- EIAC-F antibody is slow to rise and develop in about 1 month.

This test is excellent for coccidioidomycosis because it is quantitative. However, these antibodies cross-react with some other fungi (Blastomyces and Histoplasma).If the titer keeps rising, then the patient is responding poorly and the course may be fatal. If the C-F titer is dropping then the prognosis for that patient is favorable. A titer of greater than 1:128 usually indicates extensive dissemination. Life-long immunity usually follows infection with *C. immitis*. There is a much greater mortality rate in dark-skinned people (Mexicans, Filipinos, and Blacks). They are 25 times more likely to develop progressive disease and death. The reason for this is obscure.

Amphotericin B, fluconazole and itraconazole are the drugs of choice.

PARACOCCIDIOIDOMYCOSIS

This is a chronic granulomatous disease of mucous

membranes, skin, and pulmonary system. This disease occurs from the middle of Mexico (North America) to Central and South America. Most cases are reported from Brazil. The ecological niche of this organism is probably the soil. A common triad of symptoms that are seen in Latin America is pulmonary lesions, edentulous mouth, and cervical lymphadenopathy.

Prior to the recognition of this disease, patients in Latin America with paracoccidioidomycosis were often sent to TB sanitariums, just as patients with histoplasmosis were in the U.S. The organisms invade the mucous membranes of the mouth causing the teeth to fall out. White plaques are also found in the buccal mucosa, and this along with the triad are now used to clinically differentiate between TB and. This disease has a long latency period. 10-20 years may pass between infection and manifestation of the infection in the non-endemic areas of the world.

Typically, a case of paracoccidioidomycosis seen in the U.S. occurs in someone who worked in South America for some period of time and then they return to the U.S. and years later, develop this disease. The patient does not realise the importance of this past history. Almost all diagnoses of fungal diseases depend upon careful questioning and a probing history.The clinical material which should be sent to the lab for examination is sputum, biopsy material, pus, and crust from the lesions. Examination of sputum or crust from one of the lesions with KOH reveals a yeast because this is a dimorphic fungus. In contrast to the other yeasts, particularly Blastomyces, Paracoccidioides has multiple buds, a thin cell wall, and a narrow base. At 25 degrees C, the colony is a dense, white mycelium, not loose and cottony like the others.

On Sabouraud's agar it takes 2-3 weeks to grow. When cultured at 37 degrees C, it is slow growing with a white-tan, thick colony. Microscopically, these yeasts appear as described above ranging in size from 5 to 15 microns.

Histopathology

Histologically, one sees multiple buds forming a

"Captain's wheel." This is diagnostic of paracoccidioidomycosis. In this case, the mother cell is 40-50 microns in diameter and the buds are 2-5 microns in size.

Serology

The best serological test for paracoccidioidomycosis is the immunodiffusion test. It is better than 99% specific and almost 85% sensitive.

Therapy

The D.O.C. is amphotericin B. Sulphonamide-trimethoprim and ketoconazole have also been used. Presently Itraconazole appear to provide the best recovery.

SPOROTRICHOSIS

Sporotrichosis is usually a chronic infection of the cutaneous or subcutaneous tissue which tends to suppurate, ulcerate and drain. In recent years, a pulmonary disease has been seen more frequently. Occasionally, infection with *S. schenckii* may result in a mycetoma. Sporotrichosis is caused by another dimorphic fungus. The infection is also known as "rose growers disease." The ecologic niche for this organism is rose thorns, sphagnum moss, timbers and soil.

A study on the occupational distribution of sporotrichosis showed that forest employees accounted for 17% of the cases, gardeners and florists, 10%; and other soil-related occupations another 16%. Sporotrichosis occurs worldwide. Every aspect of this disease (clinical, pathology, mycology, ecology) was investigated during an epidemic of 3,000 patients in a gold mine in South Africa during the 1940's. Patient history is very important in this disease also. It is often seen in gardeners and begins with a thorn prick on the thumb.

A pustule develops and ulcerates. It infects the lymphatic system and then the disease progresses up the arm with ulceration, abscess formation, break down of the abscess with large amounts of pus followed by healing. Progression usually stops at the axilla. Clinical material to be sent to the lab may be pus, biopsy material, or sputum from pulmonary patients.

The yeast form of this fungus in tissue or in culture, can be round (6 - 8 *u*m) or fusiform.

The fusiform shape is not the usual form but if a cigar-shaped yeast is observed in tissue, it is usually diagnostic of sporotrichosis. *S. schenckii* does not stain with the usual histopathological stains. If sporotrichosis is suspected, the pathologist must be informed so he can use special stains. Histologically asteroid bodies, a tissue reaction (also known as Splendori reaction) may be seen around the yeast cell.

At 25 degrees C, this colony is white-cream and very membranous (and 5), but as it ages for 2-3 weeks it becomes black and leathery. Microscopically, the mycelium is branching, septate and very delicate, 2-3 um in diameter. The pyriform conidia, 2-4 um form a typical arrangement in groups at the end of a conidiophore called "daisies". Serologic tests are not commercially available.

The drug of choice for the cutaneous form is saturated iodides (e.g., potassium iodide) administered orally. The patient begins with 2-3 drops, 3-4/days until tolerance to the drug is built up, then the dose is increased. Potassium iodide may interact with the host immune system. For the systemic form the drug of choice is itraconazole or amphotericin B.

OPPORTUNISTIC MYCOSES

Opportunistic mycoses are infections due to fungi with low inherent virulence which means that these pathogens constitute an almost limitless number of fungi. These organisms are common in all environments.

The disease equation:

Number of organisms x Virulence = Disease

Host resistance is tilted in favour of "disease" because resistance is lowered when the host is immunocompromised. In fact, for the immunocompromised host, there is no such thing as a non-pathogenic fungus. The fungi most frequently isolated from immunocompromised patients are saprophytic (i.e. from the environment) or endogenous (a commensal). The most common species are *Candida* species, *Aspergillus*, and zygomycetes.

The upward trend in the diagnoses of opportunistic mycoses reflects increasing clinical awareness by physicians, improved clinical diagnostic procedures and better laboratory identification techniques. Another important factor contributing to the increasing incidence of infections by fungi that have not been previously known to be pathogenic has been the rise in the number of immunocompromised patients who are susceptible hosts for the most uncommon agents.

Patients with *primary* immunodeficiencies are susceptible to mycotic infections particularly when cell-mediated immunity is compromised. In addition, several types of *secondary* immunodeficiencies may be associated with an increased frequency of fungal infections.

When a fungus is isolated from an immunocompromised patient, the attending physician has to distinguish between:

- Colonization (which is of no major concern)
- Transient fungemia (often involving *C. albicans*)
- Systemic infection.

A great deal of clinical judgment is required to reach these conclusions, which imply important therapeutic decisions.

The diagnosis of opportunistic infections requires a high index of suspicion. Without this curiosity, the clinician may not consider mycotic infections in the compromised patient because:

- Patients present with atypical signs and symptoms
- The etiological agent may be considered a saprophyte or contaminant
- The systemic mycoses may occur outside the known endemic area.

Causes of immunodeficiency commonly encountered are:

- Malignancies: (Leukemias, lymphomas, Hodgkin's Disease). In one study of cancer patients, fungal septicemia and pneumonias accounted for almost a third of deaths.
- *Drug therapies*: Anti-neoplastic substances, steroids, immunosuppressive drugs.
- *Antibiotics*: Over-use or inappropriate use of antibiotics can also contribute to the development of

fungal infections by altering the normal flora of the host and facilitating fungal overgrowth or by selecting for resistant organisms.

Therapeutic procedures can predispose for fungal infections:

- Solid Organ and Bone Marrow transplantation
- Open heart surgery
- Indwelling catheters (urinary, I.V. drugs or parenteral hyperalimentation). In cases of fungemia, the contaminated catheter must be removed before starting anti-fungal therapy.
- Artificial heart valves can be colonized by a variety of infectious agents, including *Candida* species. In a case of Candida infection of an artificial heart valve, antifungal treatment is only efficient if the infected valve is replaced.
- Radiation therapy.

Other factors associated with increased frequency of mycotic infections are:

- Severe burns
- Diabetes
- Tuberculosis
- I.V. drug use
- AIDS: Virtually all AIDS patients will have a fungal infection sometime during the course of disease.

Certain fungi may be frequently associated with some of the predisposing factors listed above. However, any one of the ubiquitous saprophytes (most of which do not cause disease in immunocompetent hosts) as well as occasional pathogens may cause disease in these patients.

Biofilm Formation

It has long been recognized that in patients with a microbial infection, any artificial device such as an indwelling catheter or prosthetic valve, must be removed prior to initiating antibiotic therapy. The foreign body will act as a nidus, seeding the infection if it remains present. The exact mechanism is not clear. A biofilm is a microcolony of organisms which adhere to a surface (catheter, implant, or dead tissue) and which resist removal by fluid movement and have a decreased susceptibility to antimicrobials.

This biofilm phenomenon, which occurs on the rocks in a stream, was first recognized as a public health problem in water pipes and was regarded as a source of coliform contamination of drinking water. Recent work in clinical microbiology has shown that these organisms develop a resistance to therapy because they are contained in a matrix which acts like a tissue to and becomes a barrier to antibodies and antimicrobial agents.

CLINICAL PRESENTATION

In immunosuppressed patients, common fungal infections may have an unusual presentation because of:

Atypical signs and lesions: Malassezia furfur usually causes a rather benign and self-limited disease in normal hosts (Tinea versicolor), but in immunocompromised patients may show a rash with disseminated disease and sepsis. This organism requires long-chain fatty acids for growth. Patients receiving parenteral fat emulsions for nutrition become a walking petri plate.

Unusual Organ affinity: Candida may invade liver, heart valves; Oral thrush occurs in people who are relatively immunocompetent while esophageal candidiasis occurs in those patients who are immunologically compromised. Cryptococcus may cause pulmonary and cutaneous infections.

Infections with systemic dimorphic fungi occurring outside endemic areas. These factors complicate the diagnosis and management of these diseases.

Unusual Histopathology: Even the inflammatory reaction may be different in biopsy specimens. The normal host reaction to fungal invasion is usually pyogenic or granulomatous. In the immunodeficient host, the reaction is necrotic.

Some examples of variations from standard fungal clinical presentation, diagnosis and treatment.

Cryptococcosis: Studies show that from 10% to 30% of

AIDS patients have cryptococcal meningitis and they will require maintenance therapy with fluconazole for the remainder of their life. Fluconazole penetrates the cerebrospinal fluid

Mortality: Without treatment–100%
With treatment–20%
Relapse: Non-AIDS patients 15–20%
IDS patients –50%
With relapse there is 60% mortality.

Sporotrichosis: Co-infection with other fungi is frequent.

Coccidioidomycosis: Mycelial forms seen in tissue. Occurs in patients outside the endemic area. Patients require fluconazole or itraconazole maintenance therapy.

Histoplasmosis: All cases are disseminated.

Relapse rate is > 50% and the infection is rapidly fatal in 10% of patients. It occurs in patients outside the endemic area and they require fluconazole or itraconazole maintenance therapy.

Blastomycosis: More frequently disseminated. All patients have done very poorly.

There has been one report on 15 cases of blastomycosis in AIDS patients. Six patients (40%) had CNS involvement. Usually CNS disease only occurs in 3-10% of the patients.

- Aspergillosis
  Mortality: With amphotericin B 72%
  Without amphotericin B 90%
- Penicillium marneffei: This is a dimorphic fungus that produces a red pigment and reproduces by fission. It requires amphotericin B therapy and oral itraconazole maintenance.
- Pneumocystis carinii

This was formerly thought to be a protozoan. Presently it is believed to be a fungus.

Chapter 9

Immunology

Immunology is a broad branch of biomedical science that covers the study of all aspects of the immune system in all organisms. It deals with, among other things, the physiological functioning of the immune system in states of both health and disease; malfunctions of the immune system in immunological disorders (autoimmune diseases, hypersensitivities, immune deficiency, transplant rejection); the physical, chemical and physiological characteristics of the components of the immune system in vitro, in situ, and in vivo.

Immunology has applications in several disciplines of science, and as such is further divided. Immunology is the study of our protection from foreign macromolecules or invading organisms and our responses to them. These invaders include viruses, bacteria, protozoa or even larger parasites. In addition, we develop immune responses against our own proteins (and other molecules) in autoimmunity and against our own aberrant cells in tumor immunity.

A second line of defence is the specific or adaptive immune system which may take days to respond to a primary invasion (that is infection by an organism that has not hitherto been seen). In the specific immune system, we see the production of antibodies (soluble proteins that bind to foreign antigens) and cell-mediated responses in which specific cells recognize foreign pathogens and destroy them. In the case of viruses or tumors, this response is also vital to the recognition and destruction of virally-infected or tumorigenic cells.

The response to a second round of infection is often more rapid than to the primary infection because of the activation of memory B and T cells. We shall see how cells of the immune system interact with one another by a variety of signal molecules so that a coordinated response may be mounted. These signals may be proteins such as lymphokines which are produced by cells of the lymphoid system, cytokines and chemokines that are produced by other cells in an immune response, and which stimulate cells of the immune system.

IMMUNE SYSTEM

It is our immune system that enables us to resist infections. The immune system is composed of two major subdivisions, the innate or nonspecific immune system and the adaptive or specific immune system. The innate immune system is our first line of defence against invading organisms while the adaptive immune system acts as a second line of defence and also affords protection against re-exposure to the same pathogen.

Each of the major subdivisions of the immune system has both cellular and humoral components by which they carry out their protective function. In addition, the innate immune system also has anatomical features that function as barriers to infection. Although these two arms of the immune system have distinct functions, there is interplay between these systems (i.e., components of the innate immune system influence the adaptive immune system and vice versa). Although the innate and adaptive immune systems both function to protect against invading organisms, they differ in a number of ways.

The adaptive immune system requires some time to react to an invading organism, whereas the innate immune system includes defenses that, for the most part, are constitutively present and ready to be mobilized upon infection. Second, the adaptive immune system is antigen specific and reacts only with the organism that induced the response. In contrast, the innate system is not antigen specific and reacts equally well to a variety of organisms. Finally, the adaptive immune system demonstrates immunological memory.

It "remembers" that it has encountered an invading organism and reacts more rapidly on subsequent exposure to the same organism. In contrast, the innate immune system does not demonstrate immunological memory. All cells of the immune system have their origin in the bone marrow and they include myeloid (neutrophils, basophils, eosinpophils, macrophages and dendritic cells) and lymphoid (B lymphocyte, T lymphocyte and Natural Killer) cells, which differentiate along distinct pathways.

The myeloid progenitor (stem) cell in the bone marrow gives rise to erythrocytes, platelets, neutrophils, monocytes/macrophages and dendritic cells whereas the lymphoid progenitor (stem) cell gives rise to the NK, T cells and B cells. For T cell development the precursor T cells must migrate to the thymus where they undergo differentiation into two distinct types of T cells, the CD4+ T helper cell and the CD8+ pre-cytotoxic T cell.

Two types of T helper cells are produced in the thymus the TH1 cells, which help the CD8+ pre-cytotoxic cells to differentiate into cytotoxic T cells, and TH2 cells, which help B cells, differentiate into plasma cells, which secrete antibodies. The main function of the immune system is self/non-self discrimination. This ability to distinguish between self and non-self is necessary to protect the organism from invading pathogens and to eliminate modified or altered cells (e.g. malignant cells).

Since pathogens may replicate intracellularly (viruses and some bacteria and parasites) or extracellularly (most bacteria, fungi and parasites), different components of the immune system have evolved to protect against these different types of pathogens. It is important to remember that infection with an organism does not necessarily mean diseases, since the immune system in most cases will be able to eliminate the infection before disease occurs. Disease occurs only when the bolus of infection is high, when the virulence of the invading organism is great or when immunity is compromised.

ELEMENTS OF NON-SPECIFIC IMMUNITY

The elements of the non-specific (innate) immune system include anatomical barriers, secretory molecules and cellular components. Among the mechanical anatomical barriers are the skin and internal epithelial layers, the movement of the intestines and the oscillation of broncho-pulmonary cilia. Associated with these protective surfaces are chemical and biological agents.

Cellular Barriers to Infection

Part of the inflammatory response is the recruitment of polymorphonuclear eosinophiles and macrophages to sites of infection. These cells are the main line of defence in the non-specific immune system.

Neutrophils: Polymorphonuclear cells (PMNs, 4) are recruited to the site of infection where they phagocytose invading organisms and kill them intracellularly. In addition, PMNs contribute to collateral tissue damage that occurs during inflammation.

Macrophages: Tissue macrophages and newly recruited monocytes which differentiate into macrophages, also function in phagocytosis and intracellular killing of microorganisms. In addition, macrophages are capable of extracellular killing of infected or altered self target cells. Furthermore, macrophages contribute to tissue repair and act as antigen-presenting cells, which are required for the induction of specific immune responses.

Natural killer (NK) and lymphokine activated killer (LAK) cells – NK and LAK cells can nonspecifically kill virus infected and tumor cells.

These cells are not part of the inflammatory response but they are important in nonspecific immunity to viral infections and tumor surveillance.

Eosinophils: Eosinophils have proteins in granules that are effective in killing certain parasites.

ANATOMICAL FACTORS OF INFECTIONS

Biological Factors

The normal flora of the skin and in the gastrointestinal tract can prevent the colonization of pathogenic bacteria by secreting toxic substances or by competing with pathogenic bacteria for nutrients or attachment to cell surfaces.

Chemical Factors

Fatty acids in sweat inhibit the growth of bacteria. Lysozyme and phospholipase found in tears, saliva and nasal secretions can breakdown the cell wall of bacteria and destabilize bacterial membranes. The low pH of sweat and gastric secretions prevents growth of bacteria. Defensins (low molecular weight proteins) found in the lung and gastrointestinal tract have antimicrobial activity. Surfactants in the lung act as opsonins (substances that promote phagocytosis of particles by phagocytic cells).

Mehanical Factors

The epithelial surfaces form a physical barrier that is very impermeable to most infectious agents. Thus, the skin acts as our first line of defence against invading organisms. The desquamation of skin epithelium also helps remove bacteria and other infectious agents that have adhered to the epithelial surfaces. Movement due to cilia or peristalsis helps to keep air passages and the gastrointestinal tract free from microorganisms. The flushing action of tears and saliva helps prevent infection of the eyes and mouth. The trapping effect of mucus that lines the respiratory and gastrointestinal tract helps protect the lungs and digestive systems from infection.

Humoral Barriers to Infection

The anatomical barriers are very effective in preventing colonization of tissues by microorganisms. However, when there is damage to tissues the anatomical barriers are breached and infection may occur. Once infectious agents have penetrated tissues, another innate defence mechanism comes into play, namely acute inflammation. Humoral factors play an important role in inflammation, which is characterized by edema and the recruitment of phagocytic cells. These humoral factors are found in serum or they are formed at the site of infection.

Complement system: The complement system is the major humoral non-specific defence mechanism. Once activated complement can lead to increased vascular permeability, recruitment of phagocytic cells, and lysis and opsonization of bacteria.

Coagulation system: Depending on the severity of the tissue injury, the coagulation system may or may not be activated. Some products of the coagulation system can contribute to the non-specific defenses because of their ability to increase vascular permeability and act as chemotactic agents for phagocytic cells. In addition, some of the products of the coagulation system are directly antimicrobial. For example, beta-lysin, a protein produced by platelets during coagulation can lyse many Gram positive bacteria by acting as a cationic detergent.

Lactoferrin and transferrin: By binding iron, an essential nutrient for bacteria, these proteins limit bacterial growth.

Interferons: Interferons are proteins that can limit virus replication in cells.

Lysozyme: Lysozyme breaks down the cell wall of bacteria.

Interleukin-1 – Il-1 induces fever and the production of acute phase proteins, some of which are antimicrobial because they can opsonize bacteria.

PHAGOCYTOSIS AND INTRACELLULAR KILLING

Phagocytic Cells

Neutrophiles/Polymorphonuclear cells: PMNs are motile phagocytic cells that have lobed nuclei. They can be identified by their characteristic nucleus or by an antigen present on the cell surface called CD66. They contain two kinds of granules the contents of which are involved in the antimicrobial properties of these cells. The primary or azurophilic granules, which are abundant in young newly formed PMNs, contain cationic proteins and defensins that can kill bacteria, proteolytic enzymes like elastase, and cathepsin G to breakdown proteins, lysozyme to break down bacterial cell walls, and characteristically, myeloperoxidase, which is involved in the generation of bacteriocidal compounds. The second type of granule found in more mature PMNs is the secondary or specific granule. These contain lysozyme, NADPH oxidase components, which are involved in the generation of toxic oxygen products, and characteristically lactoferrin, an iron chelating protein and B12-binding protein.

Monocytes/Macrophages: Macrophages are phagocytic cells that have a characteristic kidney-shaped nucleus. They can be identified morphologically or by the presence of the CD14 cell surface marker. Unlike PMNs they do not contain granules but they have numerous lysosomes which have contents similar to the PNM granules.

Response of Phagocytes to Infection

Circulating PMNs and monocytes respond to danger (SOS) signals generated at the site of an infection. SOS signals include N-formyl-methionine containing peptides released by bacteria, clotting system peptides, complement products and cytokines released from tissue macrophages that have encountered bacteria in tissue. Some of the SOS signals stimulate endothelial cells near the site of the infection to express cell adhesion molecules such as ICAM-1 and selectins which bind to components on the surface of phagocytic cells and cause the phagocytes to adhere to the endothelium.

Initiation of Phagocytosis

Phagocytic cells have a variety of receptors on their cell membranes through which infectious agents bind to the cells. These include:

Complement receptors: Phagocytic cells have a receptor for the 3rd component of complement, C3b. Binding of C3b-coated bacteria to this receptor also results in enhanced phagocytosis and stimulation of the respiratory burst.

Scavenger receptors: Scavenger receptors bind a wide variety of polyanions on bacterial surfaces resulting in phagocytosis of bacteria.

Fc receptors: Bacteria with IgG antibody on their surface have the Fc region exposed and this part of the Ig molecule can bind to the receptor on phagocytes. Binding to the Fc receptor requires prior interaction of the antibody with an antigen. Binding of IgG-coated bacteria to Fc receptors results in enhanced phagocytosis and activation of the metabolic activity of phagocytes (respiratory burst).

Toll-like receptors: Phagocytes have a variety of Toll-like receptors (Pattern Recognition Receptors or PRRs) which recognize broad molecular patterns called PAMPs (pathogen associated molecular patterns) on infectious agents. Binding of infectious agents via Toll-like receptors results in phagocytosis and the release of inflammatory cytokines (IL-1, TNF-alpha and IL-6) by the phagocytes.

After attachment of a bacterium, the phagocyte begins to extend pseudopods around the bacterium. The pseudopods eventually surround the bacterium and engulf it, and the bacterium is enclosed in a phagosome.

FUNCTIONS OF COMPLEMENT

Historically, the term complement (C) was used to refer to a heat-labile serum component that was able to lyse bacteria (activity is destroyed (inactivated) by heating serum at 56 degrees C for 30 minutes). However, complement is now known to contribute to host defenses in other ways as well. Complement can opsonize bacteria for enhanced phagocytosis; it can recruit and activate various cells including polymorphonuclear cells (PMNs) and macrophages; it can participate in regulation of antibody responses and it can aid in the clearance of immune complexes and apoptotic cells. Complement can also have detrimental effects for the host; it contributes to inflammation and tissue damage and it can trigger anaphylaxis.

Complement comprises over0 different serum proteins that are produced by a variety of cells including, hepatocytes, macrophages and gut epithelial cells. Some complement proteins bind to immunoglobulins or to membrane components of cells. Others are proenzymes that, when activated, cleave one or more other complement proteins. Upon cleavage some of the complement proteins yield fragments that activate cells, increase vascular permeability or opsonize bacteria.

LECTIN PATHWAY OF COMPLEMENT ACTIVATION

It is initiated by the binding of mannose-binding lectin (MBL) to bacterial surfaces with mannose-containing polysaccharides (mannans). Binding of MBL to a pathogen results in the association of two serine proteases, MASP-1 and MASP-2 (MBL-associated serine proteases). MASP-1 and MASP-2 are similar to C1r and C1s, respectively and MBL is similar to C1q. Formation of the MBL/MASP-1/MASP-2 tri-molecular complex results in the activation of the MASPs and subsequent cleavage of C4 into C4a and C4b. The C4b fragment binds to the membrane and the C4a fragment is released into the microenvironment.

Activated MASPs also cleave C2 into C2a and C2b. C2a binds to the membrane in association with C4b and C2b is released into the microenvironment. The resulting C4bC2a complex is a C3 convertase, which cleaves C3 into C3a and C3b. C3b binds to the membrane in association with C4b and C2a and C3a is released into the microenvironment. The resulting C4bC2aC3b is a C5 convertase. The generation of C5 convertase is the end of the lectin pathway. The biological activities and the regulatory proteins of the lectin pathway are the same as those of the classical pathway.

ALTERNATIVE PATHWAY

The alternative pathway begins with the activation of C3 and requires Factors B and D and Mg^{++} cation, all present in normal serum.

Generation of C5 Convertase

Some of the C3b generated by the stabilized C3 convertase on the activator surface associates with the C3bBb complex to form a C3bBbC3b complex. This is the C5 convertase of the alternative pathway. The generation of C5 convertase is the end of the alternative pathway. The alternative pathway can be activated by many Gram-negative (most significantly,

Neisseria meningitidis and *N. gonorrhoea*), some Gram-positive bacteria and certain viruses and parasites, and results in the lysis of these organisms.

Thus, the alternative pathway of C activation provides another means of protection against certain pathogens before an antibody response is mounted. A deficiency of C3 results in an increased susceptibility to these organisms. The alternate pathway may be the more primitive pathway and the classical and lectin pathways probably developed from it.

Remember that the alternative pathway provides a means of non-specific resistance against infection without the participation of antibodies and hence provides a first line of defence against a number of infectious agents.

Many gram negative and some gram positive bacteria, certain viruses, parasites, heterologous red cells, aggregated immunoglobulins (particularly, IgA) and some other proteins (e.g. proteases, clotting pathway products) can activate the alternative pathway. One protein, cobra venom factor (CVF), has been extensively studied for its ability to activate this pathway.

Amplification Loop of C3b Formation

In serum there is low level spontaneous hydrolysis of C3 to produce C3i. Factor B binds to C3i and becomes susceptible to Factor D, which cleaves Factor B into Bb. The C3iBb complex acts as a C3 convertase and cleaves C3 into C3a and C3b. Once C3b is formed, Factor B will bind to it and becomes susceptible to cleavage by Factor D.

The resulting C3bBb complex is a C3 convertase that will continue to generate more C3b, thus amplifying C3b production. If this process continues unchecked, the result would be the consumption of all C3 in the serum. Thus, the spontaneous production of C3b is tightly controlled.

Control of the Amplification Loop

As spontaneously produced C3b binds to autologous host membranes, it interacts with DAF (decay accelerating factor), which blocks the association of Factor B with C3b thereby preventing the formation of additional C3 convertase. In addition, DAF accelerates the dissociation of Bb from C3b in C3 convertase that has already formed, thereby stopping the production of additional C3b. Some cells possess complement receptor 1 (CR1). Binding of C3b to CR1 facilitates the enzymatic degradation of C3b by Factor I.

In addition, binding of C3 convertase (C3bBb) to CR1 also dissociates Bb from the complex. Thus, in cells possessing complement receptors, CR1 also plays a role in controlling the amplification loop. Finally, Factor H can bind to C3b bound to a cell or in the in the fluid phase and facilitate the enzymatic degradation of C3b by Factor I.

Thus, the amplification loop is controlled by either blocking the formation of C3 convertase, dissociating C3 convertase, or by enzymatically digesting C3b. The importance of controlling this amplification loop is illustrated in patients with genetic deficiencies of Factor H or I. These patients have a C3 deficiency and increased susceptibility to certain infections.

Stabilization of C Convertase by Activator

When bound to an appropriate activator of the alternative pathway, C3b will bind Factor B, which is enzymatically cleaved by Factor D to produce C3 convertase (C3bBb). However, C3b is resistant to degradation by Factor I and the C3 convertase is not rapidly degraded, since it is stabilized by the activator surface. The complex is further stabilized by properdin binding to C3bBb.

Activators of the alternate pathway are components on the surface of pathogens and include: LPS of Gram-negative bacteria and the cell walls of some bacteria and yeasts. Thus, when C3b binds to an activator surface, the C3 convertase formed will be stable and continue to generate additional C3a and C3b by cleavage of C3.

BIOLOGICALLY ACTIVATION OF COMPLEMENT

Activation of complement results in the production of several biologically active molecules which contribute to resistance, anaphylaxis and inflammation.

Chemotactic Factors

C5a and MAC (C5b67) are both chemotactic. C5a is also a potent activator of neutrophils, basophils and macrophages and causes induction of adhesion molecules on vascular endothelial cells.

Opsonins

C3b and C4b in the surface of microorganisms attach to C-receptor (CR1) on phagocytic cells and promote phagocytosis.

Kinin Production

C2b generated during the classical pathway of C activation is a prokinin which becomes biologically active following enzymatic alteration by plasmin. Excess C2b production is prevented by limiting C2 activation by C1 inhibitor (C1-INH) also known as serpin which displaces C1rs from the C1qrs complex.

A genetic deficiency of C1-INH results in and overproduction of C2b and is the cause of hereditary angioneurotic edema. This condition can be treated with Danazol which promotes C1-INH production or with μ-amino caproic acid which decreases plasmin activity.

Anaphylotoxins

C4a, C3a and C5a (in increasing order of activity) are all anaphylotoxins which cause basophil/mast cell degranulation and smooth muscle contraction. Undesirable effects of these peptides are controlled by carboxypeptidase B (C3a-INA).

ANTIGENS

Antigen (Ag): A substance that reacts with the products of a specific immune response.

Hapten: A substance that is non-immunogenic but which can react with the products of a specific immune response. Haptens are small molecules which could never induce an immune response when administered by themselves but which can when coupled to a carrier molecule. Free haptens, however, can react with products of the immune response after such products have been elicited. Haptens have the property of antigenicity but not immunogenicity.

Immunogen: A substance that induces a specific immune response.

Epitope: That portion of an antigen that combines with the products of a specific immune response.

Antibody (Ab): A specific protein which is produced in response to an immunogen and which reacts with an antigen.

TYPES OF ANTIGENS

T-Dependent Antigens

T-dependent antigens are those that do not directly stimulate the production of antibody without the help of T cells. Proteins are T-dependent antigens. Structurally these antigens are characterized by a few copies of many different antigenic determinants.

T-independent Antigens

T-independent antigens are antigens which can directly stimulate the B cells to produce antibody without the requirement for T cell help In general, polysaccharides are T-independent antigens. The responses to these antigens differ from the responses to other antigens.

Properties of T-Independent Antigens

Resistance to degradation: T-independent antigens are generally more resistant to degradation and thus they persist for longer periods of time and continue to stimulate the immune system.haracterized by a few copies of many different antigenic determinants.

Polymeric structure: These antigens are characterized by the same antigenic determinant repeated many times.

Polyclonal activation of B cells: Many of these antigens can activate B cell clones specific for other antigens (polyclonal activation). T-independent antigens can be subdivided into Type 1 and Type 2 based on their ability to polyclonally activate B cells. Type 1 T-independent antigens are polyclonal activators while Type 2 are not.

Contribution of the Immunogen

Chemical Composition: In general, the more complex the substance is chemically the more immunogenic it will be. The antigenic determinants are created by the primary sequence of residues in the polymer and/or by the secondary, tertiary or quaternary structure of the molecule.

Physical form: In general particulate antigens are more immunogenic than soluble ones and denatured antigens more immunogenic than the native form.

Degradability: Antigens that are easily phagocytosed are generally more immunogenic. This is because for most antigens (T-dependant antigens) the development of an immune response requires that the antigen be phagocytosed, processed and presented to helper T cells by an antigen presenting cell (APC).

Foreignness: The immune system normally discriminates between self and non-self such that only foreign molecules are immunogenic.

Size: There is not absolute size above which a substance will be immunogenic. However, in general, the larger the molecule the more immunogenic it is likely to be.

Route: Generally the subcutaneous route is better than the intravenous or intragastric routes. The route of antigen administration can also alter the nature of the response

Adjuvants: Substances that can enhance the immune response to an immunogen are called adjuvants. The use of adjuvants, however, is often hampered by undesirable side effects such as fever and inflammation.

Dose: The dose of administration of an immunogen can influence its immunogenicity. There is a dose of antigen above or below which the immune response will not be optimal.

CHEMICAL NATURE OF IMMUNOGENS

- *Proteins:* The vast majority of immunogens are proteins. These may be pure proteins or they may be glycoproteins or lipoproteins. In general, proteins are usually very good immunogens.

- *Polysaccharides* : Pure polysaccharides and lipopolysaccharides are good immunogens.
- *Nucleic Acids:* Nucleic acids are usually poorly immunogenic. However, they may become immunogenic when single stranded or when complexed with proteins.
- *Lipids:* In general lipids are non-immunogenic, although they may be haptens.

SUPERANTIGENS

The immune system encounters a conventional T-dependent antigen, only a small fraction (1 in 10^4 -10^5) of the T cell population is able to recognize the antigen and become activated (monoclonal/oligoclonal response). However, there are some antigens which polyclonally activate a large fraction of the T cells (up to 25%). These antigens are called superantigens.

Examples of superantigens include: Staphylococcal enterotoxins (food poisoning), Staphylococcal toxic shock toxin (toxic shock syndrome), Staphylococcal exfoliating toxins (scalded skin syndrome) and Streptococcal pyrogenic exotoxins (shock). Although the bacterial superantigens are the best studied there are superantigens associated with viruses and other microorganisms as well.

HAPTEN-CARRIER CONJUGATES

Hapten-carrier conjugates are immunogenic molecules to which haptens have been covalently attached. The immunogenic molecule is called the carrier.e antigenic determinants of the carrier as well as new determinants of the hapten.

ANTIGENIC DETERMINANTS

Determinants Recognized by T Cells

Number: Although, in theory, each 8-15 residues can

constitute a separate antigenic determinant, in practice, the number of antigenic determinants per antigen is much less than what would theoretically be possible. The antigenic determinants are limited to those portions of the antigen that can bind to MHC molecules. This is why there can by differences in the responses of different individuals.

Composition: Antigenic determinants recognized by T cells are created by the primary sequence of amino acids in proteins. T cells do not recognize polysaccharide or nucleic acid antigens. This is why polysaccharides are generally T-independent antigens and proteins are generally T-dependent antigens. The determinants need not be located on the exposed surface of the antigen since recognition of the determinant by T cells requires that the antigen be proteolytically degraded into smaller peptides. Free peptides are not recognized by T cells, rather the peptides associate with molecules coded for by the major histocompatibility complex (MHC) and it is the complex of MHC molecules + peptide that is recognized by T cells.

Size: In general antigenic determinants are small and are limited to approximately 8-15 amino acids.

Determinants Recognized by B Cells

Number: Although, in theory, each 4-8 residues can constitute a separate antigenic determinant, in practice, the number of antigenic determinants per antigen is much lower than what would theoretically be possible. Usually the antigenic determinants are limited to those portions of the antigen that are accessible to antibodies (antigenic determinants are indicated in black).

Composition: Antigenic determinants recognized by B cells and the antibodies secreted by B cells are created by the primary sequence of residues in the polymer (linear or sequence determinants) and/or by the secondary, tertiary or quaternary structure of the molecule (conformational determinants).

Size: In general antigenic determinants are small and are limited to approximately 4-8 residues. (amino acids and or sugars). The combining site of an antibody will accommodate an antigenic determinant of approximately 4-8 residues.

GENETICS OF IMMUNOGLOBULINS

Glycoprotein molecules that are produced by plasma cells in response to an immunogen and which function as antibodies are called Immunoglobulins (Ig). The immunoglobulins derive their name from the finding that they migrate with globular proteins when antibody-containing serum is placed in an electrical field.

BASIC STRUCTURE OF IMMUNOGLOBULINS

Although different immunoglobulins can differ structurally they all are built from the same basic units.

Heavy and Light Chains

All immunoglobulins have a four chain structure as their basic unit. They are composed of two identical light chains (23kD) and two identical heavy chains (50-70kD).

Variable (V) and Constant (C) Regions

After the amino acid sequences of many different heavy chains and light chains were compared, it became clear that both the heavy and light chain could be divided into two regions based on variability in the amino acid sequences. These are the:

- *Light Chain:* V_L (110 amino acids) and C_L (110 amino acids)
- *Heavy Chain:* V_H (110 amino acids) and C_H (330-440 amino acids)

Hinge Region

This is the region at which the arms of the antibody molecule forms a Y. It is called the hinge region because there is some flexibility in the molecule at this point.

Domains

It is folded into globular regions each of which contains

an intra-chain disulfide bond. These regions are called domains.

- *Light Chain Domains:* V_L and C_L
- *Heavy Chain Domains:* V_H, C_{H1} - C_{H3} (or C_{H4})

Oligosaccharides

Carbohydrates are attached to the C_{H2} domain in most immunoglobulins. However, in some cases carbohydrates may also be attached at other locations.

Inter-Chain Disulfide Bonds

The heavy and light chains and the two heavy chains are held together by inter-chain disulfide bonds and by non-covalent interactions The number of inter-chain disulfide bonds varies among different immunoglobulin molecules.

Intra-chain Disulfide Binds

Within each of the polypeptide chains there are also intra-chain disulfide bonds.

FUNCTIONS OF IMMUNOGLOBULINS

ANTIGEN BINDING

Immunoglobulins bind specifically to one or a few closely related antigens. Each immunoglobulin actually binds to a specific antigenic determinant. Antigen binding by antibodies is the primary function of antibodies and can result in protection of the host. The valency of antibody refers to the number of antigenic determinants that an individual antibody molecule can bind. The valency of all antibodies is at least two and in some instances more.

EFFECTOR FUNCTIONS

Frequently the binding of an antibody to an antigen has no direct biological effect. Rather, the significant biological effects are a consequence of secondary "effector functions" of antibodies. The immunoglobulins mediate a variety of these effector functions. Usually the ability to carry out a particular effector function requires that the antibody bind to its antigen. Not every immunoglobulin will mediate all effector functions.

Binding to various cell types

Phagocytic cells, lymphocytes, platelets, mast cells, and basophils have receptors that bind immunoglobulins. This binding can activate the cells to perform some function. Some immunoglobulins also bind to receptors on placental trophoblasts, which results in transfer of the immunoglobulin across the placenta. As a result, the transferred maternal antibodies provide immunity to the fetus and newborn.

Fixation of complement

This results in lysis of cells and release of biologically active molecules.

STRUCTURE OF THE VARIABLE REGION

Hypervariable (HVR) or Complementarity Determining Regions (CDR)

Comparisons of the amino acid sequences of the variable regions of immunoglobulins show that most of the variability resides in three regions called the hypervariable regions or the complementarity determining regions. Antibodies with different specificities (i.e. different combining sites) have different complementarity determining regions while antibodies of the exact same specificity have identical complementarity determining regions (*i.e.* CDR is the antibody combining site). Complementarity determining regions are found in both the H and the L chains.

Framework Regions

The regions between the complementarity determining regions in the variable region are called the framework regions. Based on similarities and differences in the framework regions the immunoglobulin heavy and light chain variable regions can be divided into groups and subgroups. These represent the products of different variable region genes.

IMMUNOGLOBULIN FRAGMENTS

Immunoglobulin fragments produced by proteolytic digestion have proven very useful in elucidating structure/function relationships in immunoglobulins.

Fab

Digestion with papain breaks the immunoglobulin molecule in the hinge region before the H-H inter-chain disulfide bond 4. This results in the formation of two identical fragments that contain the light chain and the V_H and C_{H1} domains of the heavy chain.

Antigen Binding: These fragments were called the Fab fragments because they contained the antigen binding sites of the antibody. Each Fab fragment is monovalent whereas the original molecule was divalent. The combining site of the antibody is created by both V_H and V_L. An antibody is able to bind a particular antigenic determinant because it has a particular combination of V_H and V_L. Different combinations of a V_H and V_L result in antibodies that can bind a different antigenic determinants.

Fc

Digestion with papain also produces a fragment that contains the remainder of the two heavy chains each containing a C_{H2} and C_{H3} domain. This fragment was called Fc because it was easily crystallized.

Effector functions: The effector functions of immunoglobulins are mediated by this part of the molecule. Different functions are mediated by the different domains in this fragment. Normally the ability of an antibody to carry out an effector function requires the prior binding of an antigen; however, there are exceptions to this rule.

HUMAN IMMUNOGLOBULIN CLASSES

Immunoglobulin Classes

The immunoglobulins can be divided into five different classes, based on differences in the amino acid sequences in the constant region of the heavy chains. All immunoglobulins within a given class will have very similar heavy chain constant regions. These differences can be detected by sequence studies or more commonly by serological means (*i.e.* by the use of antibodies directed to these differences).

- IgG: Gamma heavy chains
- IgM: Mu heavy chains
- IgA: Alpha heavy chains
- IgD: Delta heavy chains
- IgE : Epsilon heavy chains

Immunoglobulin Subclasses

The classes of immunoglobulins can de divided into subclasses based on small differences in the amino acid sequences in the constant region of the heavy chains. All immunoglobulins within a subclass will have very similar heavy chain constant region amino acid sequences. Again these differences are most commonly detected by serological means.

- IgG Subclasses
  - IgG1: Gamma 1 heavy chains
  - IgG2: Gamma 2 heavy chains
  - IgG3: Gamma 3 heavy chains
  - IgG4: Gamma 4 heavy chains
- IgA Subclasses
  - IgA1: Alpha 1 heavy chains
  - IgA2: Alpha 2 heavy chains

Immunoglobulin Types

Immunoglobulins can also be classified by the type of light chain that they have. Light chain types are based on differences in the amino acid sequence in the constant region of the light chain. These differences are detected by serological means.

- Kappa light chains
- Lambda light chains

Immunoglobulin Subtypes

The light chains can also be divided into subtypes based on differences in the amino acid sequences in the constant region of the light chain.

- Lambda subtypes
  - Lambda 1
  - Lambda 2
  - Lambda 3
  - Lambda 4

Nomenclature

Immunoglobulins are named based on the class, or subclass of the heavy chain and type or subtype of light chain. Unless it is stated precisely you are to assume that all subclass, types and subtypes are present. IgG means that all subclasses and types are present.

Heterogeneity

Immunoglobulins considered as a population of molecules are normally very heterogeneous because they are composed of different classes and subclasses each of which has different types and subtypes of light chains. In addition, different immunoglobulin molecules can have different antigen binding properties because of different V_H and V_L regions.

Chapter 10

Antibodies

GENERAL CHARACTERISTICS OF ANTIBODY

Self/Non-Self Discrimination

One characteristic feature of the specific immune system is that it normally distinguishes between self and non-self and only reacts against non-self.

Specificity

A third characteristic feature of the specific immune system is that there is a high degree of specificity in its reactions. A response to a particular antigen is specific for that antigen or a few closely related antigens. N.B. These are characteristic of all specific immune responses.

Memory

A second feature of the specific immune response is that it demonstrates memory. The immune system "remembers" if it has seen an antigen before and it reacts to secondary exposures to an antigen in a manner different than after a primary exposure. Generally only an exposure to the same antigen will illicit this memory response.

ANTIBODY FORMATION

Immune elimination phase

In this phase newly synthesized antibody combines with the antigen producing antigen/antibody complexes which are phagocytosed and degraded. Antibody appears in the serum only after the immune elimination phase is over.

Equilibrium phase

The first phase is called the equilibrium or equilibration phase. During this time the Ag equilibrates between the vascular and extravascular compartments by diffusion. This is normally a rapid process. Since particulate antigens don't diffuse, they do not show this phase.

Catabolic decay phase

In this phase the host's cells and enzymes metabolize the antigen. Most of the antigen is taken up by macrophages and other phagocytic cells. The duration will depend upon the immunogen and the host.

Clearance After Secondary Injection

If there is circulating antibody in the serum injection of the antigen for a second time results in a rapid immune elimination. If the is no circulating antibody then injection of the antigen for a second time results in all three phases but the onset of the immune elimination phase is accelerated.

Kinetics of Antibody Responses to T-dependent Ag

Primary (1°) Ab Response

- Inductive, latent or lag phase: In this phase the Ag is recognized as foreign and the cells begin to proliferate and differentiate in response to the antigen. The duration of this phase will vary depending on the antigen but it is usually 5-7 days.
- Log or Exponential Phase: In this phase the Ab concentration increases exponentially as the B cells that were stimulated by the antigen differentiate into plasma cells which secrete antibody.
- Plateau or steady-state phase: In this phase Ab synthesis is balanced by Ab decay so that there in no net increase in Ab concentration.

- Decline or decay phase: In this phase the rate of Ab degradation exceeds that of Ab synthesis and the level of Ab falls. Eventually the level of Ab may reach base line levels.

Secondary (2°), Memory or Anamnestic Response

- Lag phase: In a secondary response there is a lag phase by it is normally shorter than that observed in a primary response.
- Log phase: The log phase in a secondary response is more rapid and higher Ab levels are achieved.
- Steady state phase
- Decline phase: The decline phase is not as rapid and Ab may persist for months, years or even a lifetime.

Specificity of 1° and 2° Responses

Ab elicited in response to an antigen is specific for that antigen although it may also cross react with other antigens which are structurally similar to the eliciting antigen. In general secondary responses are only elicited by the same antigen used in the primary response. However, in some instances a closely related antigen may produce a secondary response, but this is a rare exception.

Qualitative Changes in Ab during 1° and 2° Responses

Ig Class Variation

In the primary response the major class of Ab produced is IgM whereas in the secondary response it is IgG (or IgA or IgE). The antibodies that persist in the secondary response are the IgG antibodies.

Affinity

The affinity of the IgG Ab produced increases progressively during the response, particularly after low doses of antigen. This is referred to as affinity maturation. Affinity maturation is most pronounced after secondary challenge with antigen.

A second explanation for affinity maturation is that, after a class switch has occurred in the immune response, somatic mutations occur which fine tune the antibodies to be of higher affinity. There is experimental evidence for this mechanism, although it is not known how the somatic mutation mechanism is activated after exposure to antigen.

Avidity

As a consequence of increased affinity, the avidity of the antibodies increases during the response.

Cross-Reactivity

As a result of the higher affinity later in the response there is also an increase in detectible cross reactivity. An explanation for why increasing affinity results in an increase in detectible cross reactivity is illustrated by the following example. If a minimum affinity of 10^{-6} is needed to detect a reaction, early in an immune response the reaction of a cross reacting antigen with an affinity of 10^{-3} will not be detected. However, late in a response when the affinities increase 1000 fold, the reaction with both the immunizing and cross reacting antigens will be detected.

CELLULAR EVENTS DURING 1° AND 2° RESPONSES TO T-DEPENDENT AG

Primary Response

Stationary phase

As antigen is depleted, T and B cells are no longer activated. In addition, mechanisms which down regulate the immune response come into play. Furthermore, plasma cells begin to die. When the rate of antibody synthesis equals the rate of antibody decay the stationary phase is reached.

Decline phase

When no new antibody is produced because the antigen is no longer present to activate T and B cells and the residual antibody slowly is degraded, the decay phase is reached.

Lag phase

Clones of T and B cells with the appropriate antigen receptors bind antigen, become activated and begin to proliferate. The expanded clones of B cells differentiate into plasma cells which begin to secrete antibody.

Log phase

The plasma cells initially secrete IgM antibody since the C_m heavy chain gene is closest to the rearranged VDJ gene. Eventually some B cells switch from making IgM to IgG, IgA or IgE. As more B cells proliferate and differentiate into antibody secreting cells the antibody concentration increases exponentially.

Secondary Response

Not all of the T and B cells that are stimulated by antigen during primary challenge with antigen die. Some of them are long lived cells and constitute what is refer to as the memory cell pool. Both memory T cells and memory B cells are produced and memory T cells survive longer than memory B cells. Upon secondary challenge with antigen not only are virgin T and B cells activated, the memory cells are also activated and thus there is a shorter lag time in the secondary response.

Since there is an expanded clone of cells being stimulated the rate of antibody production is also increased during the log phase of antibody production and higher levels are achieved. Also, since many if not all of the memory B cells will have switched to IgG (IgA or IgE) production, IgG is produced earlier in a secondary response. Furthermore since there is an expanded clone of memory T cells which can help B cells to switch to IgG (IgA or IgE) production, the predominant class of Ig produced after secondary challenge is IgG (IgA or IgE).

Ab Response to T-independent Ag

Responses to T-independent Ag are characterized by the production of almost exclusively IgM Ab and no secondary response. Secondary exposure to the Ag results in another primary response to the Ag.

Class Switching

During an antibody response to a T-dependent antigen a switch occurs in the class of Ig produced from IgM to some other class (except IgD). Our understanding of the structure of the immunoglobulin genes, helps explain how class switching occurs.

During class switching another DNA rearrangement occurs between a switch site (S_μ) in the intron between the rearranged VDJ regions and the C_μ gene and another switch site before one of the other heavy chain constant region genes. The result of this recombination event is to bring the VDJ region close to one of the other constant region genes, thereby allowing expression of a new class of heavy chain.

Since the same VDJ gene is brought near to a different C gene and since the antibody specificity is determined by the hypervariable regions within the V region, the antibody produced after the switch occurs will have the same specificity as before. Cytokines secreted by T helper cells can cause the switch to certain isotypes.

Membrane and Secreted Immunoglobulin

The specificity of membrane immunoglobulin on a B cell and the Ig secreted by the plasma cell progeny of a B cell is the same. An understanding of how the specificity of membrane and secreted Ig from an individual B cell can be the same comes from an understanding of immunoglobulin genes.

There are two potential polyA sites in the immunoglobulin gene. One after the exon for the last heavy chain domain and the other after the exons that code for the transmembrane domains. If the first polyA site is used, the pre-mRNA is processed to produce a secreted protein. If the second polyA site is used, the pre-mRNA is processed to produce a membrane form of the immunoglobulin.

However, in all cases the same VDJ region is used and

thus the specificity of the antibody remains the same. All C regions genes have these additional membrane pieces associated with them and thus after class switching other classes of immunoglobulins can be secreted or expressed on the surface of B cells.

STRUCTURE AND FUNCTIONS

ALLOTYPES

Allotypes are antigenic determinants specified by allelic forms of the Ig genes. Allotypes represent slight differences in the amino acid sequences of heavy or light chains of different individuals. Even a single amino acid difference can give rise to an allotypic determinant, although in many cases there are several amino acid substitutions that have occurred. Allotypic differences are detected by using antibodies directed against allotypic determinants.

These antibodies can be prepared by injecting the Ig from one person into another. In practice however we obtain anti-allotype antisera from women who have had multiple pregnancies or from people who have received blood transfusions or from some patients with rheumatoid arthritis.

Occurrence

Individual allotypes are found in individual members of a species. All allotypes are not found in all members of the species. The prefix Allo means different in individuals of a species.

Human Ig Allotypes

Nomenclature: Human Ig allotypes are named on the basis of the heavy or light chain on which it is located. Thus, an allotype on a Gamma 1 heavy chain is given the name: G1m. An allotype on a Kappa light chain is given the name: Km(1)

ISOTYPES

Isotypes are antigenic determinants that characterize classes and subclasses of heavy chains and types and subtypes of light chains. If human IgM is injected into a rabbit the rabbit will recognize antigenic determinants on the heavy chain and light chain and make antibodies to them. If that antiserum is absorbed with human IgG the antibodies to the light chain determinants and any determinants in common between human IgM and IgG will be removed and the resulting antiserum will be react only with human IgM.

Indeed, the antibodies will only react with the constant region of the ¼ chain. Antibodies to the variable region are rare perhaps because only a few copies of each different variable region are represented in the IgM and thus effective immunization does not occur. The determinants that are recognized by such antibodies are called *isotypic determinants* and the antibodies to those determinants are called *anti-isotypic antibodies*. Each class, subclass, type and subtype of immunoglobulin has its unique set of isotypic determinants.

Location

Heavy chain isotypes are found on the Fc portion of the constant region of the molecule while light chain isotypes are found in the constant region.

Occurrence

Isotypes are found in ALL NORMAL individuals in the species. The prefix Iso means same in all members of the species. Some individuals with immunodeficiencies may lack one or more isotypes but normal individuals have all isotypes.

Importance

Antibodies to isotypes are used for the quantitation of Ig classes and subclasses in various diseases, in the characterization of B cell leukemia and in the diagnosis of various immunodeficiency diseases. Immunoglobulin allotypes.

IDIOTYPES (Id)

Unique antigenic determinants present on individual antibody molecules or on molecules of identical specificity.

Identical specificity means that all antibodies molecules have the exact same hypervariable regions. Antigenic determinants created by the combining site of an antibody are called idiotypes and the antibodies elicited to the idiotypes are called anti-Id antibodies.

Idiotypes are the antigenic determinants created by the hypervariable regions of an antibody and the anti-idiotypic antibodies are those directed against the hypervariable regions of an antibody.

To understand what idiotypes are, it is helpful to understand how they are detected.

DNP-BSA ⟶ Strain A ⟶ anti-DNP Ab

↓

Antibody against the combining site of anti-DNP Ab ⟵ Strain A ⟵ purified anti-DNP Ab

An antigen, in this case the hapten dinitrophenol, is injected into a mouse and antibodies (against DNP) are elicited. This antibody can be purified to homogeneity and injected into another mouse of the same strain. Most epitopes on the antibody will be seen by the second mouse's immune system as "self"; however, the epitopes that form the binding site to DNP (idiotopes - this is a term that is not often used and frequently is used interchangeably with idiotype) will be seen as foreign since the second mouse has not been injected with DNP-BSA.

The second mouse will raise antibodies only against the idiotopes of the purified anti-DNP antibody. These are therefore anti-idiotypic antibodies. Antigenic eterminants created by the hypervariable region of an antibody are idiotypes.

Location

Idiotypes are localized on the Fab fragment of the Ig molecules Specifically, they are localized at or near the hypervariable regions of the heavy and light chains. In many instances the actual antigenic determinant (i.e. idiotype) may include some of the framework residues near the hypervariable region. Idiotypes are usually determinants created by both heavy and light chain HVR's although sometimes isolated heavy and light chains will express the idiotype.

Importance

Treatment of B cell tumors: Anti-idiotypic antibodies directed against an idiotype on malignant B cells can be used to kill the cells. Killing occurs because of complement fixation or because toxic molecules are attached to the antibodies.

V region marker: Idiotypes are a useful marker for a particular variable region.

Regulation of immune responses: There is evidence that immune responses may be regulated by anti-Id antibodies directed against our own Id's.

Vaccines: In some cases anti-idiotypic antibodies actually stimulate B cells to make antibody and thus they can be used as a vaccine. This approach is being tried to immunize against highly dangerous pathogens that cannot be safely used as a vaccine.

LIGHT CHAIN GENE FAMILIES

KAPPA LIGHT CHAINS

The *kappa* light chain gene family contains only one C region gene, since there is only one type of *kappa* light chain. There are many V region genes (approximately 250) each of which has a leader exon and a V exon. In the º gene family there are several J exons located between the V and C genes. All of the exons are separated by introns.

LAMBDA LIGHT CHAINS

The *lambda* gene family is composed of 4 C region genes, one for each subtype of lamda chain, and approximately 30 V region genes. Each of the V region genes is composed of two exons, one (L) that codes for a leader region and the other (V) that codes for most of the variable region. Upstream of each of the C genes there is and additional exon called J (joining). The L, V, J and C exons are separated by introns (intervening non-coding sequences).

Gene Rearrangement and Expression

As a cell differentiates into a mature B cell that will make a light chain, there is a rearrangement of the various genes (exons) and the gene begins to be expressed.

As a cell commits to become a B cell making a light chain, there is a rearrangement of the genes at the DNA level such that one of the V genes is brought next to one of the J regions. This occurs by a recombination event which removes the intron between the V and J regions. The selection of which V gene is used is not totally random; there is some preference for the use of V genes nearest to the J regions. However, with time all V genes can be used so that all combinations of V genes and J regions can be generated.

A consequence of this DNA rearrangement is that the gene becomes transcriptionally active because a promoter (P), which is associated with the V gene, is brought close to an enhancer (E), which is located in the intron between the J and C regions. As transcription initiates from the promoter a pre-mRNA is made which contains sequences from the L, V J and C regions as well as sequences for the introns between L and V and between J and C. This pre-mRNA is processed (spliced) in the nucleus and the remaining introns are removed. The resulting mRNA has the L, V J and C exons contiguous.

The mRNA is translated in the cytoplasm and the leader is removed as the protein is transported into the lumen of the endoplasmic reticulum. The light chain is assembled with a heavy chain in the endoplasmic reticulum and the immunoglobulin is secreted via the normal route of secretory proteins. The region V region of the mature light chain is coded for by sequences in the V gene and J region and the C region by sequences in the C gene.

Heavy Chain Gene Family

In the heavy chain gene family there are many C genes, one for each class and subclass of immunoglobulin. Each of the C genes is actually composed of several exons, one for each domain and another for the hinge region. In the heavy chain gene family there are many V region genes, each composed of a leader and V exon. In addition to several J exons, the heavy chain gene family also contains several additional exons called the D (diversity) exons. All of the exons are separated by introns.

Gene Rearrangements and Expression

As a cell commits to become a B cell making a heavy chain, there are two rearrangements at the DNA level. First, one of the D regions is brought next to one of the J regions and then one of the V genes is brought next to the rearranged DJ region. This occurs by two recombination events which remove the introns between the V, D and J regions. As with the light chains the selection of the heavy chain V gene is not totally random but eventually all of the V genes can be used.

A consequence of these DNA rearrangements is that the gene becomes transcriptionally active because a promoter (P), which is associated with the V gene, is brought close to an enhancer (E), which is located in the intron between the J and C_{mu} regions. As transcription initiates from the promoter a pre-mRNA is made which contains sequences from the L, V, D, J C_{mu} and C_{delta} regions as well as sequences for the introns between L and V, between J and C_{mu}, and between C_{mu} and C_{delta}.

The pre-mRNA is processed (spliced) in the nucleus and the remaining introns, including those between the exons in the C genes, are removed. The pre-mRNA can be processed in two ways, one to bring the VDJ next to the C_{mu} gene and the other to bring the VDJ next to the C_{delta} gene. The resulting mRNAs have the L, V, D, J and C_{mu} or C_{delta} exons contiguous and will code for a *mu* and a *delta* chain, respectively.

The mRNAs are translated in the cytoplasm and the leader is removed as the protein is transported into the lumen of the endoplasmic reticulum. The heavy chain is assembled with a light chain in the endoplasmic reticulum and the immunoglobulin is secreted via the normal route of secretory proteins. The region V region of the mature heavy chain is coded for by sequences in the V gene, D region and J region and the C region by sequences in the C gene.

Mechanism of DNA Rearrangements

Flanking the V, J and D exons there are unique sequences referred to as recombination signal sequences (RSS), which function in recombination. Each RSS consists of a conserved nonamer and a conserved heptamer that are separated by either 12 or 23 base pairs (bp). The 12bp and 23 bp spaces correspond to one or two turns of the DNA helix.

Recombination only occurs between a 1 turn and a 2 turn signal. In the case of the λ light chains there is a 1 turn signal upstream of the J exon and a 2 turn signal downstream of V_{lambda}. In the case of the κ light chains there is a 1 turn signal downstream of the V_{kappa} gene and a 2 turn signal upstream of the J exon.. In the case of the heave chains there are 1 turn signals on each side of the D exon and a 2 turn signal downstream of the V gene and a 2 turn signal upstream of the J exon. Thus, this ensures that the correct recombination events will occur.

The recombination event results in the removal of the introns between V and J in the case of the light chains or between the V, D, and J in the case of the heavy chains. The recombination event is catalyzed by two proteins, Rag-1 and Rag-2. Mutations in the genes for these proteins results in a severe combined immunodeficiency disease (both T and B cells are deficient), since these proteins and the RSS are involved in generating both the B and T cell receptors for antigen.

Order of Gene Expression in Ig Gene Families

An individual B cell only produces one type of light chain and one class of heavy chain. (*N.B.* The one exception is that a mature B cell can produce both μ and δ heavy chains but the antibody specificity is the same since the same VDJ region is found on the μ and δ chains).

Since any B cell has both maternal and paternal chromosomes which code for the immunoglobulin genes there must be some orderly way in which a cell expresses its immunoglobulin genes so as to ensure that only one type of light chain and one class of heavy chain is produced.

Heavy Chain

A cell first attempts to rearrange one of its heavy chain genes; in some cells the maternal chromosome is selected and in others the paternal chromosome is selected. If the rearrangement is successful so that a heavy chain is made, then no further rearrangements occur in the heavy chain genes. If, on the other hand, the first attempt to rearrange the heavy chain genes is unsuccessful (*i.e.* no heavy chain is made), then the cell attempts to rearrange the heavy chain genes on its other chromosome. If the cell is unsuccessful in rearranging the heavy chain genes the second time, it is destined to be eliminated.

Kappa Light Chain

When a cell successfully rearranges a heavy chain gene, it then begins to rearrange one of its *kappa* light chain genes. It is a random event whether the maternal or paternal *kappa* light chain genes are selected. If the rearrangement is unsuccessful (*i.e.* it does not produce a functional *kappa* light chain), then it attempts to rearrange the *kappa* genes on the other chromosome. If a cell successfully rearranges a *kappa* light chain gene, it will be a B cell that makes an immunoglobulin with a *kappa* light chain.

Lambda Light Chain

If a cell is unsuccessful in rearranging both of its *kappa* light chain genes, it then attempts to make a *lambda* light chain. It is a random event whether the maternal or paternal *lambda* light chain genes are selected. If the rearrangement is unsuccessful (*i.e.* it does not produce a functional *lambda* light chain), then it attempts to rearrange the *lambda* genes on the other chromosome. If a cell successfully rearranges a *lambda* light chain gene, it will be a B cell that makes an immunoglobulin with a *lambda* light chain.

The orderly sequence of rearrangements in the immunoglobulin gene families explains:

- Why an individual B cell can only produce one kind

of immunoglobulin with one kind of heavy and one kind of light chain.

- Why a individual B cell can only make antibodies of one specificity.
- Why there is allelic exclusion in immunoglobulin allotypes at the level of an individual immunoglobulin molecule but co-dominant expression of allotypes in the organism as a whole.

T Cell Receptor for Antigen

T cells also have a receptor for antigen on their surfaces. This receptor is not an immunoglobulin molecule but it is composed of two different polypeptide chains which have constant and variable regions analogous to the immunoglobulins. Diversity in the T cell receptor is also generated in the same way as described for antibody diversity (*e.g.* by VJ and VDJ joining of gene segments and combinatorial association). However, no somatic mutation has been observed in T cells.

IMMUNOGLOBULINS - ANTIGEN - ANTIBODY REACTIONS AND SELECTED TESTS

NATURE OF ANTIGEN-ANTIBODY REACTIONS

Lock and Key Concept

The combining site of an antibody is located in the Fab portion of the molecule and is constructed from the hypervariable regions of the heavy and light chains. X-Ray crystallography studies of antigen-antibody interactions show that the antigenic determinant nestles in a cleft formed by the combining site of the antibody. Thus, our concept of antigen-antibody reactions is one of a key (*i.e.* the antigen) which fits into a lock (*i.e.* the antibody).

Non-Covalent Bonds

The bonds that hold the antigen to the antibody combining site are all non-covalent in nature. These include hydrogen bonds, electrostatic bonds, Van der Waals forces and hydrophobic bonds. Multiple bonding between the antigen and the antibody ensures that the antigen will be bound tightly to the antibody.

Reversibility

Since antigen-antibody reactions occur via non-covalent bonds, they are by their nature reversible.

AFFINITY AND AVIDITY

Affinity

Antibody affinity is the strength of the reaction between a single antigenic determinant and a single combining site on the antibody. It is the sum of the attractive and repulsive forces operating between the antigenic determinant and the combining site of the antibody.

Affinity is the equilibrium constant that describes the antigen-antibody reaction. Most antibodies have a high affinity for their antigens.

Avidity

Avidity is a measure of the overall strength of binding of an antigen with many antigenic determinants and multivalent antibodies. Avidity is influenced by both the valence of the antibody and the valence of the antigen. Avidity is more than the sum of the individual affinities.

To repeat, affinity refers to the strength of binding between a single antigenic determinant and an individual antibody combining site whereas avidity refers to the overall strength of binding between multivalent antigens and antibodies.

SPECIFICITY AND CROSS REACTIVITY

Specificity

Specificity refers to the ability of an individual antibody combining site to react with only one antigenic determinant or the ability of a population of antibody molecules to react with only one antigen. In general, there is a high degree of specificity in antigen-antibody reactions. Antibodies can distinguish differences in:

- The primary structure of an antigen,
- Isomeric forms of an antigen, and
- Secondary and tertiary structure of an antigen.

Cross Reactivity

Cross reactivity refers to the ability of an individual antibody combining site to react with more than one antigenic determinant or the ability of a population of antibody molecules to react with more than one antigen. Cross reactions arise because the cross reacting antigen shares an epitope in common with the immunizing antigen or because it has an epitope which is structurally similar to one on the immunizing antigen (multispecificity).

CELLS INVOLVED IN IMMUNE RESPONSES

Immunology is the study of the mechanisms that a host has evolved to rid itself of pathogens and other foreign substances.

There are two sites at which pathogens may be located:

- Extracellular sites
- Intracellular sites

Antibodies are effective against extracellular pathogens and function in three major ways:

- Neutralization
  - Antibodies may bind to bacterial toxins
  - Antibodies may bind to molecules that viruses and bacteria use to attach to cells to gain entry for infection.
- *Opsonization*: An antibody facilitates uptake by phagocytes
- *Complement activation*: Antibodies facilitate uptake by phagocytes and lyse certain bacteria

It should be noted that antibodies in each class can have different sites of action and are not equally effective in neutralization, opsonization, and complement activation. Antibodies are not particularly effective against pathogens that reside intracellularly.

CELL-MEDIATED RESPONSES

Cell-mediated responses are effective against intracellular pathogens, which reside in one of two major intracellular compartments:

- *Cytosol:* Continuous with nucleus via nuclear pores - site of all viruses and some bacteria.
- *Vesicular system:* Comprises endoplasmic reticulum, Golgi apparatus, endosomes, lysosomes, and other intracellular vesicles - site of some bacteria and some parasites

Cells harboring pathogens in the cytosol are recognized and killed by cytotoxic T cells, also known as CD8 T cells, cytolytic T cells.

Cells harboring pathogens in the vesicular system are recognized by a subpopulation of helper T cells, the Th1 T cell or inflammatory T cell which releases products called cytokines that enable the infected cell to kill the pathogen.

Pathogens may elicit both an antibody (humoral) and cell-mediated response, each of which contributes to ridding the host of the pathogen.

For example: cells with intracellular viruses can be killed by cytotoxic T cells; viruses that are free extracellularly can be neutralized and opsonized by antibody.

A humoral or cell-mediated response may by itself be insufficient to eliminate the pathogen.

For example: *Mycobacterium leprae*, an intracellular bacterium that can cause leprosy. There are two main classes of patients:

- In some, the major response is Ig production and *M. Leprae* grows abundantly in macrophages leading to gross tissue destruction and development of lepromatous leprosy which is usually fatal
- In others, little Ig is produced, but there are few live intracellular *M. leprae* because a cell-mediated

response has also occurred. Thus, disease progression is slow with a good outcome.

POPULATIONS OF T CELLS

T cells play a central role both in humoral immune (antibody) responses and cell-mediated responses. There are subpopulations of T cells that have the following functions:

- Cytotoxic Tc cells. These cells express CD8 antigen and kill cells that have pathogens in the cytosol.
- Helper Th cells. These cells express CD4 antigen and belong to two groups.
  - Inflammatory Th1 cells involved in the elimination of pathogens residing intracellularly in vesicular compartments.
  - "True" helper Th2 cells required for antibody production by B cells

Each of these "cell - cell" interactions will be discussed in more detail.

Specificity of Immune Responses

The specificity for these immune responses resides in the T cell receptor (TCR) which recognizes pathogen (antigen)-derived peptides bound to major histocompatibility complex (MHC) molecules expressed on the surface of nucleated cells. Every TCR on an individual T cell has one specificity. (*REMEMBER:* the B cell receptor that binds antigen is a membrane-bound immunoglobulin, and every Ig on an individual B cell has one specificity.

Diversity of Receptor Specificity

This is frequently referred to as the repertoire.

Historical note: The discovery and elucidation of the T cell receptor (TCR) is recent. Antibodies and their ability to bind many different antigens have been known for a long time, and two major hypotheses were advanced to explain how antibodies could be formed that bind to many antigens:

- Instructionist (template) hypothesis
- Clonal selection hypothesis

The instructionist hypothesis did not account for recognition of self vs. non-self antigens. Because of this and increasing scientific evidence, the clonal selection hypothesis has been accepted.

This hypothesis accounts for the specificity of an immune response, signals required for specific activation, lag in adaptive (specific) immune responses, and lack of an immune response against self antigens.

Self-reactive T cells are eliminated mainly in the thymus.

LYMPHOCYTE RECIRCULATION

Lymphocytes recirculate and encounter antigen in peripheral lymphoid tissues.

Since so few lymphocytes possess a receptor that can bind a given antigen (1/10,000 to 1/100,000), chances for a successful encounter with an antigen presenting cell are optimized by circulating lymphocytes through lymphoid tissues.

IMMUNITY: CONTRASTS BETWEEN NON-SPECIFIC AND SPECIFIC

- Non-specific (natural, native, innate)
  - System in place prior to exposure to antigen
  - Lacks discrimination among antigens
  - Can be enhanced after exposure to antigen through effects of cytokines
- Specific (acquired, adaptive)
  - Induced by antigen
  - Enhanced by antigen
  - Shows fine discrimination

The hallmarks of the specific immune system are memory and specificity.

- The specific immune system "remembers" each encounter with a microbe or foreign antigen, so that subsequent encounters stimulate increasingly effective defence mechanisms.
- The specific immune response amplifies the protective mechanisms of non-specific immunity, directs or focuses these mechanisms to the site of

antigen entry, and thus makes them better able to eliminate foreign antigens.

CELLS OF THE IMMUNE SYSTEM

All cell types in the immune system originate from the bone marrow.

There are two main lineages that derive from the hemopoietic stem cell:

- The lymphoid lineage
- T lymphocytes (T cells)
- B lymphocytes (B cells)
- Natural killer cells (NK cells)
- The myeloid lineage
- Monocytes, macrophages
- Langerhans cells, dendritic cells
- Megakaryocytes
- Granulocytes (eosinophils, neutrophils, basophils)

LEUKOCYTE MIGRATION AND LOCALIZATION

Productive cell interactions leading to specific immune responses occur mainly in lymph nodes and spleen. (The lymph node and spleen are referred to as secondary lymphoid tissues; bone marrow and thymus are termed primary lymphoid tissues.)presenting cells (APCs), including dendritic cells and mononuclear phagocytes (monocytes), also derive from bone marrow stem cells. These APCs enter tissues, take up antigen and transport it to the lymphoid tissues to be presented to T cells and B cells. Primed lymphocytes then migrate from the lymphoid tissues and accumulate preferentially at sites of infection and inflammation.

- T cells exiting the thymus and B cells leaving the bone marrow (naive lymphocytes, also called virgin lymphocytes) migrate from primary lymphoid tissues to the blood and thence into secondary lymphoid tissues. From there they return to the circulation and ultimately back to the secondary lymphoid tissues. This is known as lymphocyte recirculation. Approximately 1-2% of the lymphocyte pool recirculates each hour and optimizes the opportunities for antigen-specific lymphocytes to come into contact with antigen in the secondary lymphoid tissues.

- Lymphocytes activated by an encounter with antigen move from the spleen or lymph node and travel to other tissues. For example, activated T cells express new cell surface molecules that enable them to bind to peripheral vascular endothelium and thence to egress to extravascular spaces. Primed and activated are synonymous terms.
- Antigen presenting cells may pick up antigen and migrate to the secondary lymphoid tissues. The Langerhans cell of the skin is a classic example.

RESPONSE TO ANTIGEN: PROCESSING AND PRESENTATION

MHC RESTRICTION AND ROLE OF THE THYMUS

Review Of B And T Cell Receptors For Antigen

B cells and T cells recognize different substances as antigens and in a different form. The B cell uses cell surface-bound immunoglobulin as a receptor and the specificity of that receptor is the same as the immunoglobulin that it is able to secrete after activation. B cells recognize the following antigens in soluble form:

- Proteins (both conformational determinants and determinants exposed by denaturation or proteolysis)
- Nucleic acids
- Polysaccharides
- Some lipids
- Small chemicals (haptens)

In contrast, the overwhelming majority of antigens for T cells are proteins, and these must be fragmented and recognized in association with MHC products expressed on the surface of nucleated cells, not in soluble form. T cells are grouped functionally according to the class of MHC molecules that associate with the peptide fragments of protein: helper T cells recognize only those peptides associated with class II MHC molecules, and cytolytic T cells recognize only those peptides associated with class I MHC molecules.

Antigen Processing And Presentation

Antigen processing and presentation are processes that occur within a cell that result in fragmentation (proteolysis) of proteins, association of the fragments with MHC molecules, and expression of the peptide-MHC molecules at the cell surface where they can be recognized by the T cell receptor on a T cell. However, the path leading to the association of protein fragments with MHC molecules differs for class I and class II MHC. MHC class I molecules present degradation products derived from intracellular (endogenous) proteins in the cytosol. MHC class II molecules present fragments derived from extracellular (exogenous) proteins that are located in an intracellular compartment.

Antigen processing and presentation in cells expressing class I MHC: All nucleated cells express class I MHC.Proteins are fragmented in the cytosol by proteosomes (a complex of proteins having proteolytic activity) or by other proteases. The fragments are then transported across the membrane of the endoplasmic reticulum by transporter proteins.

(The transporter proteins and some components of the proteosome have their genes in the MHC complex). Synthesis and assembly of class I heavy chain and beta$_2$ microglobulin occurs in the endoplasmic reticulum. Within the endoplasmic reticulum, the MHC class I heavy chain, beta$_2$microglobulin and peptide form a stable complex that is transported to the cell surface.

Antigen processing and presentation in cells expressing class II MHC: Whereas all nucleated cells express class I MHC, only a limited group of cells express class II MHC, which includes the antigen presenting cells (APC). The principal APC are macrophages, dendritic cells (Langerhans cells), and B cells, and the expression of class II MHC molecules is either constitutive or inducible, especially by interferon-gamma in the case of macrophages. Exogenous proteins taken in by endocytosis are fragmented by proteases in an endosome. The alpha and beta chains of MHC class II, along with an invariant chain, are synthesized, assembled in the endoplasmic reticulum, and transported through the Golgi and trans-Golgi apparatus to reach the endosome, where the invariant chain is digested, and the peptide fragments from the exogenous protein are able to associate with the class II MHC molecules, which are finally transported to the cell surface.

Other points concerning antigen processing and presentation. One way of rationalizing the development of two different pathways is that each ultimately stimulates the population of T cells that is most effective in eliminating that type of antigen.

Viruses replicate within nucleated cells in the cytosol and produce endogenous antigens that can associate with class I MHC. By killing these infected cells, cytolytic T cells help to control the spread of the virus.

Bacteria mainly reside and replicate extracellularly. By being taken up and fragmented inside cells as exogenous antigens that can associate with class II MHC molecules, helper Th2 T cells can be activated to assist B cells to make antibody against bacteria, which limits the growth of these organisms.

Some bacteria grow intracellularly inside the vesicles of cells like macrophages. Inflammatory Th1 T cells help to activate macrophages to kill the intracellular bacteria.

- Fragments of self, as well as non-self, proteins associate with MHC molecules of both classes and are expressed at the cell surface.
- Which protein fragments bind is a function of the chemical nature of the groove for that specific MHC molecule

Self MHC Restriction

In order for a T cell to recognize and respond to a foreign protein antigen, it must recognize the MHC on the presenting cell as self MHC. This is termed self MHC restriction. Helper T cells recognize antigen in context of class II self MHC.

Cytolytic T cells recognize antigen in context of class I self MHC. The process whereby T cells become restricted to recognizing self MHC molecules occurs in the thymus.

INTERACTIONS IN IMMUNE RESPONSES

Hapten-Carrier Effect

Historically one of the major findings was that T cells and B cells are required in order to produce antibody to a complex protein. A major contribution to our understanding of this process came from studies on the formation of anti-hapten antibodies.

Recall that a hapten injected by itself cannot elicit an antibody response. Rather antibodies against haptens require that the hapten be conjugated to a protein (sometimes termed a carrier).

These studies with hapten-carrier established that:

- Th cells recognize the carrier, and B cells recognize hapten.
- There must be cooperation between hapten-specific B cells and protein (carrier)-specific helper T cells.
- Interaction between the hapten-specific B cell and the carrier-specific helper T cell are class II self MHC-restricted. The helper T cell cooperates only with B cells that express class II MHC molecules recognized as self by the T cells.

B Cells as Antigen Presenting Cells

B cells occupy a unique position in immune responses because they express immunoglobulin (Ig) and class II MHC molecules on their cell surface.

They therefore are capable of producing antibody having the same specificity as that expressed by their immunoglobulin receptor; in addition they can function as an antigen presenting cell.

In terms of the hapten-carrier protein findings, the mechanism is thought to be the following: the hapten is recognized by the Ig receptor, the hapten-carrier brought into the B cell, processed, and peptide fragments of the carrier protein presented to a helper T cell. Activation of the T cell results in the production of cytokines that enable the hapten-specific B cell to become activated to produce soluble anti-hapten antibodies.

Note that there are multiple signals delivered to the B cells in this model of Th cell-B cell interaction. As was the case for activation of T cells where the signal derived from the TCR recognition of a peptide-MHC molecule was by itself insufficient for T cell activation, so too for the B cell.

Binding of an antigen to the immunoglobulin receptor delivers one signal to the B cell, but that is insufficient. Second signals delivered by costimulatory molecules are required; the most important of these is CD40L on the T cell that binds to CD40 on the B cell to initiate delivery of a second signal.

Antigen-presenting cells (e.g. dendritic cells) present processed antigen to virgin T cells, thereby priming them. B cells also process the antigen and present it to the T cells. They then receive signals from the T cells that cause them to divide and differentiate. Some B cells form antibody-forming cells while a few form B memory cells.

Extension of this Model to Complex Protein Antigens (T-dependent Antigens)

The same mechanism described above can cover all multideterminant complex protein antigens that require helper T cells. These antigens are referred to as thymus-dependent antigens. If one determinant is recognized by B cells (analogous to the hapten) and the same or different determinant is recognized by the helper T cells (analogous to the carrier), the same model applies.

B cells in Secondary Responses

As a consequence of a primary response, many memory B cells are created. These carry a high-affinity receptor, Ig, which allows them to bind and present antigen at much lower concentrations than is required for macrophages or dendritic cells.

IMMUNIZATION

Immunization is the means of providing specific protection against most common and damaging pathogens. Specific immunity can be acquired either by passive or by active immunization and both modes of immunization can occur by natural or artificial means.

PASSIVE IMMUNITY

Immunity can be acquired, without the immune system being challenged with an antigen. This is done by transfer of serum or gamma-globulins from an immune donor to a non-immune individual. Alternatively, immune cells from an immunized individual may be used to transfer immunity. Passive immunity may be acquired naturally or artificially.

Naturally Acquired Passive Immunity

Immunity is transferred from mother to fetus through placental transfer of IgG or colostral transfer of IgA.

Artificially Acquired Passive Immunity

Immunity is often artificially transferred by injection with gamma-globulins from other individuals or gamma-globulin from an immune animal.

Passive transfer of immunity with immune globulins or gamma-globulins is practiced in numerous acute situations of infections (diphtheria, tetanus, measles, rabies, etc.), poisoning (insects, reptiles, botulism), and as a prophylactic measure (hypogammaglobulinemia). In these situations, gamma-globulins of human origin are preferable although specific antibodies raised in other species are effective and used in some cases (poisoning, diphtheria, tetanus, gas gangrene, botulism).

While this form of immunization has the advantage of providing immediate protection, heterologous gamma-globulins are effective for only a short duration and often result in pathological complications (serum sickness) and anaphylaxis. Homologous immunoglobulins carry the risk of transmitting hepatitis and HIV.

ACTIVE IMMUNITY

This refers to immunity produced by the body following exposure to antigens.

Naturally Acquired Active Immunity

Exposure to different pathogens leads to sub-clinical or clinical infections which result in a protective immune response against these pathogens.

Artificially Acquired Active Immunity

Immunization may be achieved by administering live or dead pathogens or their components. Vaccines used for active immunization consist of live (attenuated) organisms, killed whole organisms, microbial components or secreted toxins (which have been detoxified).

The first live vaccine was cowpox virus introduced by Edward Jenner as a vaccine for smallpox; however, variolation, innoculation using pus from a patient with a mild case of smallpox has been in use for over a thousand years.

Live vaccines are used against a number of viral infections {polio (Sabin vaccine), measles, mumps, rubella, chicken pox, hepatitis A, yellow fever, *etc.*} The only example of live bacterial vaccine is one against tuberculosis (*Mycobacterium bovis*: BCG).

Anti-idiotype antibodies are also under trial. Similarly, DNA vaccines and immunodominant peptides (recognized by the MHC molecules) are under investigation, particularly for protection against viral diseases.

The protective immunity conferred by a vaccine may be lifelong (measles, mumps, rubella, small pox, tuberculosis, yellow fever, etc.) or may last as little as six months (cholera).

The primary immunization may be given at the age of 2 - 3 months (diphtheria, pertussis, tetanus, polio), or 13 - 15 months (mumps, measles, rubella). A number of other vaccines are licensed for use in the US and are recommended for individuals or groups at risk.

Active immunization may cause fever, malaise and

discomfort. Some vaccine may also cause joint pains or arthritis (rubella), convulsions, sometimes fatal (pertussis), or neurological disorders (influenza). Allergies to eggs may develop as a consequence of viral vaccines produced in eggs (measles, mumps, influenza, yellow fever).

Index